How We Learn

ALSO BY STANISLAS DEHAENE

Consciousness and the Brain:
Deciphering How the Brain Codes Our Thoughts

Reading in the Brain:
The New Science of How We Read

The Number Sense:
How the Mind Creates Mathematics

How We Learn

Why Brains Learn Better Than
Any Machine . . . for Now

STANISLAS DEHAENE

VIKING

VIKING
An imprint of Penguin Random House LLC
penguinrandomhouse.com

ISBN 9780525559887 (hardcover)
ISBN 9780525559894 (ebook)

Printed in the United States of America
10 9 8 7 6 5 4 3 2 1

Book design by Daniel Lagin

For Aurore, who was born this year,

and for all those who once were babies.

Begin by making a more careful study of your pupils, for it is clear that you know nothing about them.

Jean-Jacques Rousseau, *Emile, or On Education* (1762)

This is a strange and amazing fact: we know every nook and cranny of the human body, we have catalogued every animal on the planet, we have described and baptized every blade of grass, but we have left psychological techniques to their empiricism for centuries, as if they were of lesser importance than those of the healer, the breeder or the farmer.

Jean Piaget, "La pédagogie moderne" (1949)

If we don't know how we learn, how on earth do we know how to teach?

L. Rafael Reif, president of MIT (March 23, 2017)

CONTENTS

Part Three

The Four Pillars of Learning

INTRODUCTION

IN SEPTEMBER 2009, AN EXTRAORDINARY CHILD FORCED ME TO DRASTI-
cally revise my ideas about learning. I was visiting the Sarah Hospital in Brasilia,
a neurological rehabilitation center with a white architecture inspired by
Oscar Niemeyer, with which my laboratory has collaborated for about ten
years. The director, Lucia Braga, asked me to meet one of her patients, Felipe,
a young boy only seven years old, who had spent more than half his life in a
hospital bed. She explained to me how, at the age of four, he had been shot in
the street—unfortunately not such a rare event in Brazil. The stray bullet had
severed his spinal cord, thus rendering him almost completely paralyzed (tet-
raparetic). It also destroyed the visual areas of his brain: he was fully blind. To
help him breathe, an opening was made in his trachea, at the base of his neck.
And for over three years, he had been living in a hospital room, locked within
the coffin of his inert body.

In the corridor leading to his room, I remember bracing myself at the
thought of having to face a broken child. And then I meet . . . Felipe, a lovely
little boy like any other seven-year-old—talkative, full of life, and curious
about everything. He speaks flawlessly with an extensive vocabulary and asks
me mischievous questions about French words. I learn that he has always been
passionate about languages and never misses an opportunity to enrich his

trilingual vocabulary (he speaks Portuguese, English, and Spanish). Although he is blind and bedridden, he escapes into his imagination by writing his own novels, and the hospital team has encouraged him in this path. In a few months, he learned to dictate his stories to an assistant, then write them himself using a special keyboard connected to a computer and sound card. The pediatricians and speech therapists take turns at his bedside, transforming his writings into real, tactile books with embossed illustrations that he proudly sweeps with his fingers, using the little sense of touch that he has left. His stories speak of heroes and heroines, mountains and lakes that he will never see, but that he dreams of like any other little boy.

Meeting with Felipe deeply moved me, and also persuaded me to take a closer look at what is probably the greatest talent of our brain: the ability to learn. Here was a child whose very existence poses a challenge to neuroscience. How do our brain's cognitive faculties resist such a radical upheaval of their environment? Why could Felipe and I share the same thoughts, given our extraordinarily different sensory experiences? How do different human brains converge on the same concepts, almost regardless of how and when they learn them?

Many neuroscientists are empiricists: together, with the English Enlightenment philosopher John Locke (1632–1704), they presume that the brain simply draws its knowledge from its environment. In this view, the main property of cortical circuits is their plasticity, their ability to adapt to their inputs. And, indeed, nerve cells possess a remarkable ability to constantly adjust their synapses according to the signals they receive. Yet if this were the brain's main drive, my little Felipe, deprived of visual and motor inputs, should have become a profoundly limited person. By what miracle did he manage to develop strictly normal cognitive abilities?

Felipe's case is by no means unique. Everybody knows the story of Helen Keller (1880–1968) and Marie Heurtin (1885–1921), both of whom were born deaf and blind and yet, after years of grueling social isolation, learned sign language and ultimately became brilliant thinkers and writers.[1] Throughout these pages, we will meet many other individuals who, I hope, will radically alter your views on learning. One of them is Emmanuel Giroux, who has been blind since the age of eleven but became a top-notch mathematician. Paraphrasing the fox in Antoine de Saint-Exupéry's *The Little Prince* (1943), Giroux confidently states: "In geometry, what is essential is invisible to the eye. It is only with the mind that you can see well." How does this blind man manage to swiftly navigate within the abstract spaces of algebraic geometry, manipulating planes, spheres, and volumes without ever seeing them? We will discover that he uses the same brain circuits as other mathematicians, but that his visual cortex, far from remaining inactive, has actually repurposed itself to do math.

I will also introduce you to Nico, a young painter who, while visiting the Marmottan Museum in Paris, managed to make an excellent copy of Monet's famous painting *Impression, Sunrise* (see figure 1 in the color insert). What is so exceptional about this? Nothing, besides the fact that he accomplished it with only a single hemisphere, his left one—the right half of his brain was almost fully removed at the age of three! Nico's brain learned to squeeze all his talents into half a brain: speech, writing, and reading, as usual, but drawing and painting too, which are generally thought to be functions of the right hemisphere, and also computer science and even wheelchair fencing, a sport in which he has reached the rank of champion in Spain. Forget everything you were told about the respective roles of both hemispheres, because Nico's life proves that anyone can become a creative and talented artist without a right hemisphere! Cerebral plasticity seems to work miracles.

We will also visit the infamous orphanages of Bucharest where children were left from birth in quasi-abandon—and yet, years later, some of them, adopted before the age of one or two, have had almost normal school experiences.

All these examples illustrate the extraordinary resilience of the human brain: even major trauma, such as blindness, the loss of a hemisphere, or social isolation, cannot extinguish the spark of learning. Language, reading, mathematics, artistic creation: all these unique talents of the human species, which

no other primate possesses, can resist massive injuries, such as the removal of
a hemisphere or the loss of sight and motor skills. Learning is a vital principle,
and the human brain has an enormous capacity for plasticity—to change itself,
to adapt. Yet we will also discover dramatic counterexamples, where learning
seems to freeze and remain powerless. Consider pure alexia, the inability to
read a single word. I have personally studied several adults, all of whom were
excellent readers, who had a tiny stroke restricted to a minuscule brain area
that rendered them incapable of deciphering words as simple as "dog" or "mat."
I remember a brilliant trilingual woman, a faithful reader of the French news-
paper *Le Monde*, who was deeply sorrowed at the fact that, after her brain injury,
every page of the daily press looked like Hebrew. Her determination to relearn
to read was at least as strong as the stroke that she had suffered was severe.
However, after two years of perseverance, her reading level still did not
exceed that of a kindergartner: it took her several seconds to read a single
word, letter by letter, and she still stumbled on every word. Why couldn't she
learn? And why do some children, who suffer from dyslexia, dyscalculia, or
dyspraxia, show a similar radical hopelessness in acquiring reading, calculating, or writing while others surf smoothly through those fields?

Brain plasticity almost seems temperamental: sometimes it overcomes
massive difficulties, and other times it leaves children and adults who are
otherwise highly motivated and intelligent with debilitating disabilities. Does
it depend on particular circuits? Do these circuits lose their plasticity over the
years? Can plasticity be reopened? What are the rules that govern it? How can
the brain be so effective from birth and throughout a child's youth? What algorithms allow our brain circuits to form a representation of the world? Would
understanding them help us learn better and faster? Could we draw inspiration from them in order to build more efficient machines, artificial intelligences that would ultimately imitate us or even surpass us? These are some
of the questions that this book attempts to answer, in a radically multidisci-
plinary manner, drawing on recent scientific discoveries in cognitive science
and neuroscience, but also in artificial intelligence and education.

WHY LEARN?

Why do we have to learn in the first place? The very existence of the capacity to learn raises questions. Wouldn't it be better for our children to immediately know how to speak and think, right from day one, like Athena, who, according to legend, emerged into the world from Zeus's skull, fully grown and armed, as she let out her war cry? Why aren't we born pre-wired, with pre-programmed software and exactly the pre-loaded knowledge necessary to our survival? In the Darwinian struggle for life, shouldn't an animal who is born mature, with more knowledge than others, end up winning and spreading its genes? Why did evolution invent learning in the first place?

My answer is simple: a complete pre-wiring of the brain is neither possible nor desirable. Impossible, really? Yes, because if our DNA had to specify all the details of our knowledge, it simply would not have the necessary storage capacity. Our twenty-three chromosomes contain three billion pairs of the "letters" A, C, G, T—the molecules adenine, cytosine, guanine, and thymine. How much information does that represent? Information is measured in bits: a binary decision, 0 or 1. Since each of the four letters of the genome codes for two bits (we can code them as 00, 01, 10, and 11), our DNA therefore contains a total of six billion bits. Remember, however, that in today's computers, we count in bytes, which are sequences of eight bits. The human genome can thus be reduced to about 750 megabytes—the contents of an old-fashioned CD-ROM or a small USB key! And this basic calculation does not even take into account the many redundancies that abound in our DNA.

From this modest amount of information, inherited from millions of years of evolution, our genome, initially confined to a single fertilized egg, manages to set up the whole body plan—every molecule of every cell in our liver, kidneys, muscles, and, of course, our brain: eighty-six billion neurons, a thousand trillion connections. . . . How could our genome possibly specify each one of them? Assuming that each of our nerve connections encodes only one bit, which is certainly an underestimate, the capacity of our brain is on the order of one hundred terabytes (about 10^{15} bits), or a hundred thousand times more than the information in our genome. We are faced with a paradox: the fantastic palace that is our brain contains a hundred thousand times more detail than the architect's blueprints that are used to build it! I see only one explanation: the structural frame of the palace is built following the architect's guidelines (our genome), while the details are left to the project manager, who can adapt the blueprints to the terrain (the environment). Pre-wiring a human brain in all its detail would be strictly impossible, which is why learning is needed to supplement the work of genes.

This simple bookkeeping argument, however, fails to explain why learning is so universally widespread in the animal world. Even simple organisms devoid of any cortex, such as earthworms, fruit flies, and sea cucumbers, learn many of their behaviors. Take the little worm called the "nematode," or *C. elegans*. In the past twenty years, this millimeter-size animal became a laboratory star, in part because its architecture is under strong genetic determinism and can be analyzed down to the smallest detail. Most individual specimens have exactly 959 cells, including 302 neurons, whose connections are all known and reproducible. And yet it learns.[2] Researchers initially considered it as a kind of robot just able to swim back and forth, but they later realized that it possesses at least two forms of learning: habituation and association. Habituation refers to an organism's capacity to adapt to the repeated presence of a stimulus (for example, a molecule in the water in which the animal lives) and eventually cease to respond to it. Association, on the other hand, consists of discovering and remembering what aspects of the environment predict sources of food or danger. The nematode worm is a champion of association: it can remember, for instance, which tastes, smells, or temperature levels were previously associated with food (bacteria) or with a repellent molecule (the smell of garlic) and use this information to choose an optimal path through its environment.

With such a small number of neurons, the worm's behavior could have been fully pre-wired. However, it is not. The reason is that it is highly advantageous, indeed indispensable for its survival, to adapt to the specific environment

in which it is born. Even two genetically identical organisms will not necessarily encounter the same ecosystem. In the case of the nematode, the ability
to quickly adjust its behavior to the density, chemistry, and temperature of
the place in which it lands allows it to be more efficient. More generally,
every animal must quickly adapt to the unpredictable conditions of its current existence. Natural selection, Darwin's remarkably efficient algorithm,
can certainly succeed in adapting each organism to its ecological niche, but
it does so at an appallingly slow rate. Whole generations must die, due to lack
of proper adaptation, before a favorable mutation can increase the species'
chance of survival. The ability to learn, on the other hand, acts much faster—
it can change behavior within the span of a few minutes, which is the very
quintessence of learning: being able to adapt to unpredictable conditions as
quickly as possible.

This is why learning evolved. Over time, the animals that possessed even
a rudimentary capacity to learn had a better chance of surviving than those
with fixed behaviors—and they were more likely to pass their genome (now
including genetically driven learning algorithms) on to the next generation. In
this manner, natural selection favored the emergence of learning. The evolutionary algorithm discovered a good trick: it is useful to let certain parameters
of the body change rapidly in order to adjust to the most volatile aspects of the
environment.

Naturally, several aspects of the physical world are strictly invariable:
gravitation is universal; the propagation of light and sound does not change
overnight; and that is why we do not have to learn how to grow ears, eyes,
or the labyrinths that, in our vestibular system, keep track of our body's
acceleration—all these properties are genetically hardwired. However, many
other parameters, such as the spacing of our two eyes, the weight and length
of our limbs, or the pitch of our voice, all vary, and this is why our brain must
adapt to them. As we shall see, our brains are the result of a compromise—we
inherit, from our long evolutionary history, a great deal of innate circuitry
(coding for all the broad intuitive categories into which we subdivide the

world: images, sounds, movements, objects, animals, people . . .) but also, perhaps, to an even greater extent, some highly sophisticated learning algorithm that can refine those early skills according to our experience.

HOMO DOCENS

If I had to sum up, in one word, the singular talents of our species, I would answer with "learning." We are not simply *Homo sapiens*, but *Homo docens*—the species that teaches itself. Most of what we know about the world was not given to us by our genes: we had to learn it from our environment or from those around us. No other animal has managed to change its ecological niche so radically, moving from the African savanna to deserts, mountains, islands, polar ice caps, cave dwellings, cities, and even outer space, all within a few thousand years. Learning has fueled it all. From making fire and designing stone tools to agriculture, exploration, and atomic fission, the story of humanity is one of constant self-reinvention. At the root of all these accomplishments lies one secret: the extraordinary ability of our brain to formulate hypotheses and select those that fit with our environment.

Learning is the triumph of our species. In our brain, billions of parameters are free to adapt to our environment, our language, our culture, our parents, or our food. . . . These parameters are carefully chosen: over the course of evolution, the Darwinian algorithm carefully delineated which brain circuits should be pre-wired and which should be left open to the environment. In our species, the contribution of learning is particularly large since our childhood extends over many more years than it does for other mammals. And because we possess a unique knack for language and mathematics, our learning device is able to navigate vast spaces of hypotheses that recombine into potentially infinite sets—even if they are always grounded in fixed and invariable foundations inherited from our evolution.

More recently, humanity discovered that it could increase this remarkable ability even further with the help of an institution: the classroom. Pedagogy is an exclusive privilege of our species: no other animal actively teaches its offspring by setting aside specific time to monitor their progress, difficulties, and errors. The invention of the school, an institution which systematizes the informal education present in all human societies, has vastly increased our brain potential. We have discovered that we can take advantage of the exuberant plasticity of the child brain to instill in it a maximum amount of information and talent. Over centuries, our school system has continued to improve in efficiency, starting earlier and earlier in childhood and now lasting for fifteen years or more. Increasing numbers of brains benefit from higher education. Universities are neural refineries where our brain circuits acquire their best talents.

Education is the main accelerator of our brain. It is not difficult to justify its presence in the top spots in government spending: without it, our cortical circuits would remain diamonds in the rough. The complexity of our society owes its existence to the multiple improvements that education brings to our cortex: reading, writing, calculation, algebra, music, a sense of time and space, a refinement of memory. . . . Did you know, for example, that the short-term memory of a literate person, the number of syllables she can repeat, is almost double that of an adult who never attended school and remained illiterate? Or that IQ increases by several points for each additional year of education and literacy?

LEARNING TO LEARN

Education magnifies the already considerable faculties of our brain—but could it perform even better? At school and at work, we constantly tinker with our brain's learning algorithms, yet we do so intuitively, without paying attention to how to learn. No one has ever explained to us the rules by which our brain memorizes and understands or, on the contrary, forgets and makes mistakes. It truly is a pity, because the scientific knowledge is extensive. An excellent website, put together by the British Education Endowment Foundation (EEF),[3] lists the most successful educational interventions – and it gives a very high ranking to the teaching of metacognition (knowing the powers and limits of one's own brain). Learning to learn is arguably the most important factor for academic success.

Fortunately, we now know a lot about how learning works. Thirty years of research, at the boundaries of computer science, neurobiology, and cognitive psychology, have largely elucidated the algorithms that our brain uses, the circuits involved, the factors that modulate their efficacy, and the reasons why they are uniquely efficient in humans. In this book, I will discuss all those points in turn. When you close this book, I hope you will know much more about your own learning processes. It seems fundamental, to me, that every child and every adult realize the full potential of his or her own brain and also, of course, its limits. Contemporary cognitive science, through the systematic dissection of our mental algorithms and brain mechanisms, gives new meaning to the famous Socratic adage "Know thyself." Today, the point is no longer just to sharpen our introspection, but to understand the subtle neuronal mechanics that generate our thoughts, in an attempt to use them in optimal accordance with our needs, goals, and desires.

The emerging science of how we learn is, of course, of special relevance to all those for whom learning is a professional activity: teachers and educators. I am deeply convinced that one cannot properly teach without possessing, implicitly or explicitly, a mental model of what is going on in the minds of the learners. What sort of intuitions do they start with? What steps do they have to take in order to move forward? What factors can help them develop their skills?

While cognitive neuroscience does not have all the answers, we begin to understand that all children start off life with a similar brain architecture—a *Homo sapiens* brain, radically different from that of other apes. I am not denying, of course, that our brains vary: the quirks of our genomes, as well as the whimsies of early brain development, grant us slightly different strengths and learning speeds. However, the basic circuitry is the same in all of us, as is the organization of our learning algorithms. There are therefore fundamental principles that any teacher must respect in order to be most effective. In this book, we will see many examples. All young children share abstract intuitions in the domains of language, arithmetic, logic, and probability, thus providing a foundation on which higher education must be grounded. And all learners benefit from focused attention, active engagement, error feedback, and a cycle of daily rehearsal and nightly consolidation—I call these factors the "four pillars" of learning, because, as we shall see, they lie at the foundation of the universal human learning algorithm present in all our brains, children and adults alike.

At the same time, our brains do exhibit individual variations, and in some extreme cases, a pathology can appear. The reality of developmental pathologies, such as dyslexia, dyscalculia, dyspraxia, and attention disorders, is no longer a subject of doubt. Fortunately, as we increasingly understand the common architecture from which these quirks arise, we also discover that simple strategies exist to detect and compensate for them. One of the goals of this book is to spread this growing scientific knowledge, so that every teacher, and also every parent, can adopt an optimal teaching strategy. While children vary dramatically in *what* they know, they still share the same learning algorithms. Thus, the pedagogical tricks that work best with all children are also those that tend to be the most efficient for children with learning disabilities—they must be applied only with greater focus, patience, systematicity, and tolerance to error.

And the latter point is crucial: while error feedback is essential, many children lose confidence and curiosity because their errors are punished rather than corrected. In schools worldwide, error feedback is often synonymous with punishment and stigmatization—and later in this book I will have much to say about the role of school grades in perpetuating this confusion. Negative emotions crush our brain's learning potential, whereas providing the brain with a fear-free environment may reopen the gates of neuronal plasticity. There will be no progress in education without simultaneously considering the emotional and cognitive facets of our brain—in today's cognitive neuroscience, both are considered key ingredients of the learning cocktail.

THE CHALLENGE OF MACHINES

Today, human intelligence faces a new challenge: we are no longer the only champions of learning. In all fields of knowledge, learning algorithms are challenging our species' unique status. Thanks to them, smartphones can now recognize faces and voices, transcribe speech, translate foreign languages, control machines, and even play chess or Go—much better than we can. Machine learning has become a billion-dollar industry that is increasingly inspired by our brains. How do these artificial algorithms work? Can their principles help us understand what learning is? Are they already able to imitate our brains, or do they still have a long way to go?

While the current advances in computer science are fascinating, their limits are evident. Conventional deep learning algorithms mimic only a small part of our brain's functioning, the one that, I argue, corresponds to the first stages of sensory processing, the first two or three hundred milliseconds during which our brain operates in an unconscious manner. This type of processing is in no way superficial: in a fraction of a second, our brain can recognize a face or a word, put it in context, understand it, and even integrate it into a small sentence. . . . The limitation, however, is that the process remains strictly bottom-up, without any real capacity for reflection. Only in the subsequent stages, which are much slower, more conscious, and more reflective, does our brain manage to deploy all its abilities of reasoning, inference, and flexibility—features that today's machines are still far from matching. Even the most advanced computer architectures fall short of any human infant's ability to build abstract models of the world.

Even within their fields of expertise—for example, the rapid recognition of shapes—modern-day algorithms encounter a second problem: they are much less effective than our brain. The state of the art in machine learning involves running millions, even billions, of training attempts on computers. Indeed, machine learning has become virtually synonymous with big data: without massive data sets, algorithms have a hard time extracting abstract knowledge that generalizes to new situations. In other words, they do not make the best use of data.

In this contest, the infant brain wins hands down: babies do not need more than one or two repetitions to learn a new word. Their brain makes the most of extremely scarce data, a competence that still eludes today's computers. Neuronal learning algorithms often come close to optimal computation: they manage to extract the true essence from the slightest observation. If computer scientists hope to achieve the same performance in machines, they will have to draw inspiration from the many learning tricks that evolution integrated into our brain: attention, for example, which allows us to select and amplify relevant information; or sleep, an algorithm by which our brain synthesizes what it learned on previous days. New machines with these properties are beginning to emerge, and their performance is constantly improving—they will undoubtedly compete with our brains in the near future.

According to an emerging theory, the reason that our brain is still superior to machines is that it acts as a statistician. By constantly attending to probabilities and uncertainties, it optimizes its ability to learn. During its evolution, our brain seems to have acquired sophisticated algorithms that constantly keep track of the uncertainty associated with what it has learned—and such a systematic attention to probabilities is, in a precise mathematical sense, the optimal way to make the most of each piece of information.[4]

Recent experimental data support this hypothesis. Even babies understand probabilities: from birth, they seem to be deeply embedded in their brain circuits. Children act like little budding scientists: their brains teem with hypotheses, which resemble scientific theories that their experiences put to the test. Reasoning with probabilities, in a largely unconscious manner, is deeply inscribed in the logic of our learning. It allows any of us to gradually reject false hypotheses and retain only the theories that make sense of the data. And, unlike other animal species, humans seem to use this sense of probabilities to acquire scientific theories from the outside world. Only *Homo sapiens* manages to systematically generate abstract symbolic thoughts and to update their plausibility in the face of new observations.

Innovative computer algorithms are beginning to incorporate this new vision of learning. They are called "Bayesian," after the Reverend Thomas

Bayes (1702–61), who outlined the rudiments of this theory as early as the eighteenth century. My hunch is that Bayesian algorithms will revolutionize machine learning—indeed, we will see that they are already able to extract abstract information with an efficiency close to that of a human scientist.

Our journey into the contemporary science of learning is a three-part trip.

In the first part, entitled "What Is Learning?", we start by defining what it means for humans or animals—or indeed any algorithm or machine—to learn something. The idea is simple: to learn is to progressively form, in silicon and neural circuits alike, an internal model of the outside world. When I walk around a new town, I form a mental map of its layout—a miniature model of its streets and passageways. Likewise, a child who is learning to ride a bike is shaping, in her neural circuits, an unconscious simulation of how the actions on the pedals and handlebars affect the bike's stability. Similarly, a computer algorithm learning to recognize faces is acquiring template models of the various possible shapes of eyes, noses, mouths, and their combinations.

But how do we set up the proper mental model? As we shall see, the learner's mind can be likened to a giant machine with millions of tunable parameters whose settings collectively define what is learned (for instance, where the streets are likely to be in our mental map of the neighborhood). In the brain, the parameters are synapses, the connections between neurons, which can vary in strength; in most present-day computers, they are the tunable weights or probabilities that specify the strength of each tenable hypothesis. Learning, in both brains and machines, thus requires searching for an optimal combination of parameters that, together, define the mental model in every detail. In this sense, learning is a massive search problem—and in order to understand how learning works in the human brain, it greatly helps to examine how learning algorithms operate in present-day computers.

By comparing the performance of computer algorithms with those of the brain, *in silico* versus *in vivo*, we will progressively get a sharper picture of what learning means at the brain level. For sure, mathematicians and computer scientists haven't managed to design learning algorithms as powerful as the human brain—yet. But they are beginning to home in on a theory of the optimal learning algorithm that any system should use if it aims for the greatest efficiency. According to this theory, the best learner operates as a scientist who makes rational use of probabilities and statistics. A new model emerges: that of the brain as a statistician, of cerebral circuits as computing with probabilities. This theory specifies a clear division of labor between nature and nurture: the genes first set up vast spaces of a priori hypotheses—and the environment then *selects* the hypotheses which best match the external world. The set of hypotheses is genetically specified; their selection is experience-dependent.

Does this theory correspond to how the brain works? And how is learning implemented in our biological circuits? What changes in our brains when we acquire a novel competence? In the second section, "How Our Brain Learns," we will turn to psychology and neuroscience. I will focus on babies, who are genuine learning machines without rivals. Recent data show that infants are indeed the budding statisticians predicted by the theory. Their remarkable intuition in the fields of language, geometry, numbers, and statistics confirms that they are anything but a blank slate, a tabula rasa. From birth, children's brain circuits are already organized and project hypotheses onto the outside world. But they also have a considerable margin of plasticity, which is reflected in the brain's perpetual effervescence of synaptic changes. Within this statistical machine, nature and nurture, far from opposing each other, join forces. The result is a structured yet plastic system with an unmatched ability to repair itself in the face of brain injury and to recycle its brain circuits in order to acquire skills unanticipated by evolution, such as reading or mathematics.

In the third part, "The Four Pillars of Learning," I detail some of the tricks that make our brain the most effective learning device known today. Four essential mechanisms, or "pillars," massively modulate our ability to learn. The first is attention: a set of neural circuits that select, amplify, and propagate the signals that we view as relevant—multiplying their impact in our memory a hundred fold. My second pillar is active engagement: a passive organism learns almost nothing, because learning requires an active generation of hypotheses, with motivation and curiosity. The third pillar, and the flip side to active engage-ment, is error feedback: whenever we are surprised because the world violates our expectations, error signals spread throughout our brain. They correct our mental models, eliminate inappropriate hypotheses, and stabilize the most accurate ones. Finally, the fourth pillar is consolidation: over time, our brain compiles what it has acquired and transfers it into long-term memory, thus freeing neural resources for further learning. Repetition plays an essential role in this consolidation process. Even sleep, far from being a period of inac-tivity, is a privileged moment during which the brain revisits its past states, at a faster pace, and recodes the knowledge acquired during the day.

These four pillars are universal: babies, children, and adults of all ages continually deploy them whenever they exercise their ability to learn. This is why we should all learn to master them—it is how we can learn to learn. In the conclusion, I will come back to the practical consequences of these scien-tific advances. Changing our practices at school, at home, or at work is not necessarily as complicated as we think. Very simple ideas about play, curios-ity, socialization, concentration, and sleep can augment what is already our brain's greatest talent: learning.

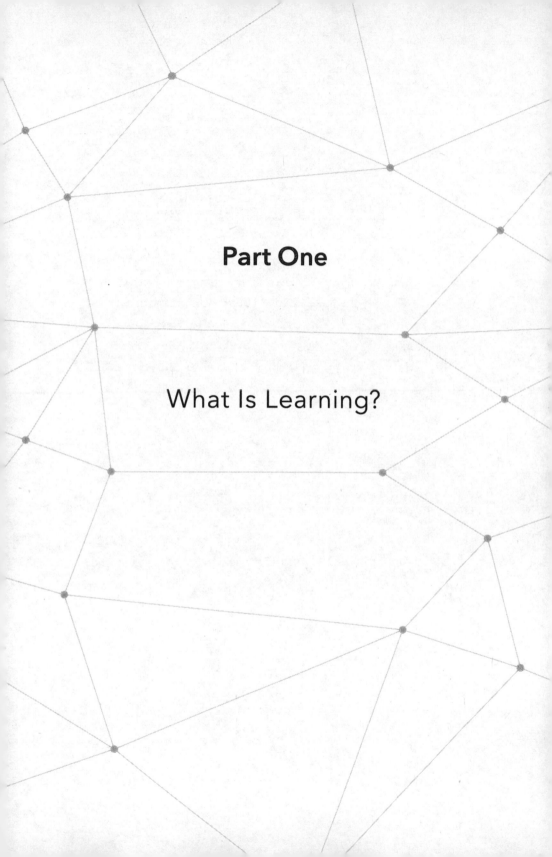

Part One

What Is Learning?

At its core, intelligence can be viewed as a process that converts unstructured information into useful and actionable knowledge.

Demis Hassabis,
founder of the AI company DeepMind (2017)

What is learning? In many Latin languages, *learning* has the same root as *apprehending: apprendre* in French, *aprender* in Spanish and Portuguese. . . . Indeed, learning is grasping a fragment of reality, catching it, and bringing it inside our brains. In cognitive science, we say that learning consists of forming an internal model of the world. Through learning, the raw data that strikes our senses turns into refined ideas, abstract enough to be reused in a new context—smaller-scale models of reality.

In the following pages, we will review what artificial intelligence and cognitive science have taught us about how such internal models emerge, in both brains and machines. How does the representation of information change when we learn? How can we understand it at a level that is common to any organism, human, animal, or machine? By reviewing the various tricks that engineers have designed to allow machines to learn, we will progressively conjure up a sharper picture of the amazing computations that infants must perform as they learn to see, speak, and write. In fact, as we shall see, the infant brain keeps the upper hand: despite their successes, current learning algorithms capture only a fraction of the abilities of the human brain. Understanding exactly where the machine learning metaphor breaks down, and where even an infant's brain still surpasses the most powerful computer, we will delineate exactly what "learning" means.

CHAPTER 1

Seven Definitions of Learning

WHAT DOES "LEARNING" MEAN? MY FIRST AND MOST GENERAL DEFINI-tion is the following: to learn is to form an internal model of the external world.

You may not be aware of it, but your brain has acquired thousands of internal models of the outside world. Metaphorically speaking, they are like miniature mock-ups more or less faithful to the reality they represent. We all have in our brains, for example, a mental map of our neighborhood and our home—all we have to do is close our eyes and envision them with our thoughts. Obviously, none of us were born with this mental map—we had to acquire it through learning.

The richness of these mental models, which are, for the most part, unconscious, exceeds our imagination. For example, you possess a vast mental model of the English language, which allows you to understand the words you are reading right now and guess that *plastovski* is not an English word, whereas *swoon* and *wistful* are, and *dragostan* could be. Your brain also includes several models of your body: it constantly uses them to map the position of your limbs and to direct them while maintaining your balance. Other mental models encode your knowledge of objects and your interactions with them: knowing how to hold a pen, write, or ride a bike. Others even represent the minds of others: you possess a vast mental catalog of people who are close to you, their appearances, their voices, their tastes, and their quirks.

These mental models can generate hyper-realistic simulations of the universe around us. Did you ever notice that your brain sometimes projects the most authentic virtual reality shows, in which you can walk, move, dance, visit new places, have brilliant conversations, or feel strong emotions? These are your dreams! It is fascinating to realize that all the thoughts that come to us in our dreams, however complex, are simply the product of our free-running internal models of the world.

But we also dream up reality when awake: our brain constantly projects hypotheses and interpretative frameworks on the outside world. This is because, unbeknownst to us, every image that appears on our retina is ambiguous—whenever we see a plate, for instance, the image is compatible with an infinite number of ellipses. If we see the plate as round, even though the raw sense data picture it as an oval, it is because our brain supplies additional data: it has learned that the round shape is the most likely interpretation. Behind the scenes, our sensory areas ceaselessly compute with probabilities, and only the most likely model makes it into our consciousness. It is the brain's projections that ultimately give meaning to the flow of data that reaches us from our senses. In the absence of an internal model, raw sensory inputs would remain meaningless.

Learning allows our brain to grasp a fragment of reality that it had previously missed and to use it to build a new model of the world. It can be a part of external reality, as when we learn history, botany, or the map of a city, but our brain also learns to map the reality internal to our bodies, as when we learn to coordinate our actions and concentrate our thoughts in order to play the violin. In both cases, our brain *internalizes* a new aspect of reality: it adjusts its circuits to appropriate a domain that it had not mastered before.

Such adjustments, of course, have to be pretty clever. The power of learning lies in its ability to adjust to the external world and to correct for errors—but how does the brain of the learner "know" how to update its internal model when, say, it gets lost in its neighborhood, falls from its bike, loses a game of chess, or misspells the word *ecstasy*? We will now review seven key ideas that lie at the heart of present-day machine-learning algorithms and that may apply equally well to our brains—seven different definitions of what "learning" means.

LEARNING IS ADJUSTING THE PARAMETERS OF A MENTAL MODEL

Adjusting a mental model is sometimes very simple. How, for example, do we reach out to an object that we see? In the seventeenth century, René Descartes (1596–1650) had already guessed that our nervous system must contain processing loops that transform visual inputs into muscular commands (see the figure on the next page). You can experience this for yourself: try grabbing an object while wearing somebody else's glasses, preferably someone who is very nearsighted. Even better, if you can, get a hold of prisms that shift your vision a dozen degrees to the left and try to catch the object.[1] You will see that your first attempt is completely off: because of the prisms, your hand reaches to the right of the object that you are aiming for. Gradually, you adjust your movements to the left. Through successive trial and error, your gestures become more and more precise, as your brain learns to correct the offset of your eyes. Now take off the glasses and grab the object: you'll be surprised to see that your hand goes to the wrong location, now way too far to the left!

So, what happened? During this brief learning period, your brain adjusted its internal model of vision. A parameter of this model, one that corresponds to the offset between the visual scene and the orientation of your body, was set to a new value. During this recalibration process, which works by trial and error, what your brain did can be likened to what a hunter does in order to adjust his rifle's viewfinder: he takes a test shot, then uses it to adjust his scope, thus progressively shooting more and more accurately. This type of learning can be very fast: a few trials are enough to correct the gap between vision and

Adjusting a single parameter: the vision-to-action offset

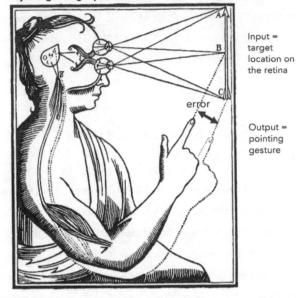

Input =
target
location on
the retina

Output =
pointing
gesture

Adjusting millions of parameters: the connections that support vision

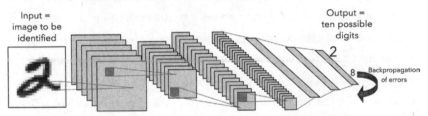

Input =
image to be
identified

Output =
ten possible
digits

Backpropagation
of errors

What is learning? To learn is to adjust the parameters of an internal model. Learning to aim with one's finger, for example, consists of setting the offset between vision and action: each aiming error provides useful information that allows one to reduce the gap. In artificial neural networks, although the number of settings is much larger, the logic is the same. Recognizing a character requires the fine-tuning of millions of connections. Again, each error—here, the incorrect activation of the output "8"—can be back-propagated and used to adjust the values of the connections, thus improving performance on the next test.

action. However, the new parameter setting is not compatible with the old one—hence the systematic error we all make when we remove the prisms and return to normal vision.

Undeniably, this type of learning is a little particular, because it requires the adjustment of only a single parameter (viewing angle). Most of our learning is much more elaborate and requires adjusting tens, hundreds, or even thousands of millions of parameters (every synapse in the relevant brain circuit). The principle, however, is always the same: it boils down to searching, among myriad possible settings of the internal model, for those that best correspond to the state of the external world.

An infant is born in Tokyo. Over the next two or three years, its internal model of language will have to adjust to the characteristics of the Japanese language. This baby's brain is like a machine with millions of settings at each level. Some of these settings, at the auditory level, determine which inventory of consonants and vowels is used in Japanese and the rules that allow them to be combined. A baby born into a Japanese family must discover which phonemes make up Japanese words and where to place the boundaries between those sounds. One of the parameters, for example, concerns the distinction between the sounds /R/ and /L/: this is a crucial contrast in English, but not in Japanese, which makes no distinction between Bill Clinton's election and his erection.... Each baby must thus fix a set of parameters that collectively specify which categories of speech sounds are relevant for his or her native language.

A similar learning procedure is duplicated at each level, from sound patterns to vocabulary, grammar, and meaning. The brain is organized as a hierarchy of models of reality, each nested inside the next like Russian dolls— and learning means using the incoming data to set the parameters at every level of this hierarchy. Let's consider a high-level example: the acquisition of grammatical rules. Another key difference which the baby must learn, between Japanese and English, concerns the order of words. In a canonical sentence with a subject, a verb, and a direct object, the English language first states the subject, then the verb, and finally its object: "John + eats + an apple." In Japanese, on the other hand, the most common order is subject, then object, then verb: "John + an apple + eats." What is remarkable is that the order is also reversed for prepositions (which logically become post-positions), possessives, and many other parts of speech. The sentence "My uncle wants to work in Boston," thus becomes mumbo jumbo worthy of Yoda from Star Wars: "Uncle my, Boston in, work wants"—which makes perfect sense to a Japanese speaker.

Fascinatingly, these reversals are not independent of one another. Linguists think that they arise from the setting of a single parameter called the "head position": the defining word of a phrase, its head, is always placed first in English (<u>in</u> Paris, <u>my</u> uncle, <u>wants to</u> live), but last in Japanese (Paris <u>in</u>, uncle <u>my</u>, live <u>wants</u>). This binary parameter distinguishes many languages, even some that are not historically linked (the Navajo language, for example, follows the same rules as Japanese). In order to learn English or Japanese, one of the things that a child must figure out is how to set the head position parameter in his internal language model.

LEARNING IS EXPLOITING A COMBINATORIAL EXPLOSION

Can language learning really be reduced to the setting of some parameters? If this seems hard to believe, it is because we are unable to fathom the extraordinary number of possibilities that open up as soon as we increase the number of adjustable parameters. This is called the "combinatorial explosion"—the exponential increase that occurs when you combine even a small number of possibilities. Suppose that the grammar of the world's languages can be described by about fifty binary parameters, as some linguists postulate. This yields 2^{50} combinations, which are over one million billion possible languages, or 1 followed by fifteen zeros! The syntactic rules of the world's three thousand languages easily fit into this gigantic space. However, in our brain, there aren't just fifty adjustable parameters, but an astoundingly larger number: eighty-six billion neurons, each with about ten thousand synaptic contacts whose strength can vary. The space of mental representations that opens up is practically infinite.

Human languages heavily exploit these combinations at all levels. Consider, for instance, the mental lexicon: the set of words that we know and whose model we carry around with us. Each of us has learned about fifty thousand words with the most diverse meanings. This seems like a huge lexicon, but we manage to acquire it in about a decade because we can decompose the learning problem. Indeed, considering that these fifty thousand words are on average two syllables, each consisting of about three phonemes, taken from the forty-four phonemes in English, the binary coding of all these words requires less than two million elementary binary choices ("bits," whose value is 0 or 1). In other words, all our knowledge of the dictionary would fit in a small 250-kilobyte computer file (each byte comprising eight bits).

This mental lexicon could be compressed to an even smaller size if we took into account the many redundancies that govern words. Drawing six letters at random, like "xfdrga," does not generate an English word. Real words are composed of a pyramid of syllables that are assembled according to strict rules. And this is true at all levels: sentences are regular collections of words, which are regular collections of syllables, which are regular collections of phonemes. The combinations are both vast (because one chooses among several tens or hundreds of elements) and bounded (because only certain combinations are allowed). To learn a language is to discover the parameters that govern these combinations at all levels.

In summary, the human brain breaks down the problem of learning by creating a hierarchical, multilevel model. This is particularly obvious in the case of language, from elementary sounds to the whole sentence or even discourse—but the same principle of hierarchical decomposition is reproduced in all sensory systems. Some brain areas capture low-level patterns: they see the world through a very small temporal and spatial window, thus analyzing the smallest patterns. For example, in the primary visual area, the first region of the cortex to receive visual inputs, each neuron analyzes only a very small portion of the retina. It sees the world through a pinhole and, as a result, discovers very low-level regularities, such as the presence of a moving oblique line. Millions of neurons do the same work at different points in the retina, and their outputs become the inputs of the next level, which thus detects "regularities of regularities," and so on and so forth. At each level, the scale broadens: the brain seeks regularities on increasingly vast scales, in both time and space. From this hierarchy emerges the ability to detect increasingly complex objects or concepts: a line, a finger, a hand, an arm, a human body . . . no, wait, two, there are two people facing each other, a handshake. . . . It is the first Trump-Macron encounter!

LEARNING IS MINIMIZING ERRORS

The computer algorithms that we call "artificial neural networks" are directly inspired by the hierarchical organization of the cortex. Like the cortex, they contain a pyramid of successive layers, each of which attempts to discover deeper regularities than the previous one. Because these consecutive layers organize the incoming data in deeper and deeper ways, they are also called "deep networks." Each layer, by itself, is capable of discovering only an extremely simple part of the external reality (mathematicians speak of a linearly separable problem, i.e., each neuron can separate that data into only two categories, A and B, by drawing a straight line through them). Assemble many of these layers, however, and you get an extremely powerful learning device, capable of discovering complex structures and adjusting to very diverse problems. Today's artificial neural networks, which take advantage of the advances in computer chips, are also deep, in the sense that they contain dozens of successive layers. These layers become increasingly insightful and capable of identifying abstract properties the further away they are from the sensory input.

Let's take the example of the LeNet algorithm, created by the French pioneer of neural networks, Yann LeCun (see figure 2 in the color insert).[2] As early as the 1990s, this neural network achieved remarkable performance in the recognition of handwritten characters. For years, Canada Post used it to automatically process handwritten postal codes. How does it work? The algorithm receives the image of a written character as an input, in the form of pixels, and it proposes, as an output, a tentative interpretation: one out of the ten possible digits or twenty-six letters. The artificial network contains a hierarchy of processing units that look a bit like neurons and form successive layers. The first layers are connected directly with the image: they apply simple filters that recognize lines and curve fragments. The layers higher up in the hierarchy, however, contain wider and more complex filters. Higher-level units can therefore learn to recognize larger and larger portions of the image: the curve of a 2, the loop of an O, or the parallel lines of a Z . . . until we reach, at the output level, artificial neurons that respond to a character regardless of its position, font, or case. All these properties are not imposed by a programmer: they result entirely from the millions of connections that link the units. These connections, once adjusted by an automated algorithm, define the filter that each neuron applies to its inputs: their settings explain why one neuron responds to the number 2 and another to the number 3.

How are these millions of connections adjusted? Just as in the case of prism glasses! On each trial, the network gives a tentative answer, is told whether it made an error, and adjusts its parameters to try to reduce this error on the next trial. Every wrong answer provides valuable information. With its sign (like a gesture too far to the right or too far to the left), the error tells the system what it should have done in order to succeed. By going back to the source of the error, the machine discovers how the parameters should have been set to avoid the mistake.

Let's revisit the example of the hunter adjusting his rifle's scope. The learning procedure is elementary. The hunter shoots and finds he's aimed five centimeters too far to the right. He now has essential information, both on the amplitude (five centimeters) and on the sign of the error (too far to the right). This information allows him to correct his shot. If he is a bit clever, he can infer in which direction to make the correction: if the bullet has deflected to the right, he should shift the scope one hair to the left. Even if he's not that astute, he can casually try a different aim and test whether, if he turns the scope to the right, the offset increases or decreases. In this manner, through trial and error, the hunter can progressively discover which adjustment reduces the size of the gap between his intended target and his actual shot.

In modifying his sight to maximize his accuracy, our brave hunter is applying a learning algorithm without even knowing it. He is implicitly calculating what mathematicians call the "derivative," or gradient, of the system, and is using the "gradient descent algorithm": he learns to move his rifle's viewfinder in the most efficient direction, the one that reduces the probability of making a mistake.

Most artificial neural networks used in present-day artificial intelligence, despite their millions of inputs, outputs, and adjustable parameters, operate just like our proverbial hunter: they observe their errors and use them to adjust their internal state in the direction that they feel is best able to reduce the errors. In many cases, such learning is tightly guided. We tell the network exactly which response it should have activated at the output ("it is a 1, not a 7"), and we know precisely in which direction to adjust the parameters if they lead to an error (a mathematical calculation makes it possible to know exactly which connections to modify when the network activates the output "7" too often in response to an image of the number 1). In machine learning parlance, this situation is known as "supervised learning" (because someone, who can be likened to a supervisor, knows the correct answer that the system must give) and "error backpropagation" (because error signals are sent back into the network in order to modify its parameters). The procedure is simple: I try an answer, I am told what I should have answered, I measure my error, and I adjust my parameters to reduce it. At each step, I make only a small correction in the right direction. That's why such computer-based learning can be incredibly slow: learning a complex activity, like playing *Tetris*, requires applying this recipe thousands, millions, even billions of times. In a space that includes a multitude of adjustable parameters, it can take a long time to discover the optimal setting for every nut and bolt.

The very first artificial neural networks, in the 1980s, were already operating on this principle of gradual error correction. Advances in computing have now made it possible to extend this idea to gigantic neural networks, which include hundreds of millions of adjustable connections. These deep neural networks are composed of a succession of stages, each of which adapts to the problem at hand. For example, figure 4 in the color insert shows the GoogLeNet system, derived from the LeNet architecture first proposed by LeCun and which won one of the most important international image recognition competitions. Exposed to billions of images, this system learned to separate them into one thousand distinct categories, such as faces, landscapes, boats, cars, dogs, insects, flowers, road signs, and so forth. Each level of its hierarchy has become attuned to a useful aspect of reality: low-level units selectively respond to lines or textures, but the higher you go up in the hierarchy, the more neurons have learned to respond to complex features, such as geometric shapes (circles, curves, stars . . .), parts of objects (a pants pocket, a car door handle, a pair of eyes . . .), or even whole objects (buildings, faces, spiders . . .).[3]

By trying to minimize errors, the gradient descent algorithm discovered that these forms are the most useful for categorizing images. But if the same network had been exposed to book passages or sheet music, it would have adjusted in a different way and learned to recognize letters, notes, or whichever shapes recur in the new environment. Figure 3 in the color insert, for example, shows how a network of this type self-organizes to recognize thousands of handwritten digits.[4] At the lowest level, the data are mixed: some images are superficially similar but should ultimately be distinguished (think of a 3 and an 8), and conversely, some images that look very different must ultimately be placed in the same bin (think of the many versions of the digit 8, with the top loop open or closed, etc.). At each stage, the artificial neural network progresses in abstraction until all instances of the same character are correctly grouped together. Through the error reduction procedure, it has discovered a hierarchy of features most relevant to the problem of recognizing handwritten digits. Indeed, it is quite remarkable that, simply by correcting one's errors, it is possible to discover a whole set of clues appropriate to the problem at hand.

Today, the concept of learning by error backpropagation remains at the heart of many computer applications. This is the workhorse that lies behind your smartphone's ability to recognize your voice, or your smart car's emerging perception of pedestrians and road signs—and it is therefore very likely that our brain uses one version of it or the other. However, error backpropagation comes in various flavors. The field of artificial intelligence has made tremendous advances in thirty years, and researchers have discovered many tricks that facilitate learning. We will now review them—as we shall see, they also tell us a lot about ourselves and the way we learn.

LEARNING IS EXPLORING THE SPACE OF POSSIBILITIES

One of the problems with the error correction procedure I just described is that it can get stuck on a set of parameters that is not the best. Imagine a golf ball rolling on the green, always along the line of the steepest slope: it may get stuck in a small depression in the ground, preventing it from reaching the lowest point of the whole landscape, the absolute optimum. Similarly, the gradient descent algorithm sometimes gets stuck at a point that it cannot exit. This is called a "local minimum": a well in parameter space, a trap from which the learning algorithm cannot escape because it seems impossible to do better. At this moment, learning gets stuck, because all changes seem counterproductive: each of them increases the error rate. The system feels that it has learned all it can. It remains blind to the presence of much better settings, perhaps only a few steps away in parameter space. The gradient descent algorithm does not "see" them because it refuses to go up the hump in order to go back down the other side of the dip. Shortsighted, it ventures only a small distance from its starting point and may therefore miss out on better but distant configurations.

Does the problem seem too abstract to you? Think about a concrete situation: You go shopping at a food market, where you spend some time looking for the cheapest products. You walk down an aisle, pass the first seller (who seems overpriced), avoid the second (who is always very expensive), and finally stop at the third stand, which seems much cheaper than the previous ones. But who's to say that one aisle over, or perhaps even in the next town, the prices would not be even more enticing? Focusing on the best *local* price does not guarantee finding the *global* minimum.

Frequently confronted with this difficulty, computer scientists employ a panoply of tricks. Most of them consist of introducing a bit of randomness in the search for the best parameters. The idea is simple: instead of looking in only one aisle of the market, take a step at random; and instead of letting the golf ball roll gently down the slope, give it a shake, thus reducing its chance of getting stuck in a trough. On occasion, stochastic search algorithms try a distant and partially random setting, so that if a better solution is within reach, they have a chance of finding it. In practice, one can introduce some degree of randomness in various ways: setting or updating the parameters at random, diversifying the order of the examples, adding some noise to the data, or using only a random fraction of the connections—all these ideas improve the robustness of learning.

Some machine learning algorithms also get their inspiration from the Darwinian algorithm that governs the evolution of species: during parameter optimization, they introduce mutations and random crossings of previously discovered solutions. As in biology, the rate of these mutations must be carefully controlled in order to explore new solutions without wasting too much time in hazardous attempts.

Another algorithm is inspired by blacksmith forges, where craftspeople have learned to optimize the properties of metal by "annealing" it. Applied when one wants to forge an exceptionally strong sword, the method of annealing consists of heating the metal several times, at lower and lower temperatures, to increase the chance that the atoms arrange themselves in a regular configuration. The process has now been transposed to computer science: the simulated annealing algorithm introduces random changes in the parameters, but with a virtual "temperature" that gradually decreases. The probability of a chance event is high at the beginning but steadily declines until the system is frozen in an optimal setting.

Computer scientists have found all these tricks to be remarkably effective—so perhaps it should be no surprise that, in the course of evolution, some of them were internalized in our brains. Random exploration, stochastic curiosity, and noisy neuronal firing all play an essential role in learning for *Homo sapiens*. Whether we are playing rock, paper, scissors; improvising on a jazz theme; or exploring the possible solutions to a math problem, randomness is an essential ingredient of a solution. As we shall see, whenever children go into learning mode—that is, when they play—they explore dozens of possibilities with a good dose of randomness. And during the night, their brains continue juggling ideas until they hit upon one that best explains what they experienced during the day. In the third section of this book, I will come back to what we know about the semi-random algorithm that governs the extraordinary curiosity of children—and the rare adults who have managed to keep a child's mind.

LEARNING IS OPTIMIZING A REWARD FUNCTION

Remember LeCun's LeNet system, which recognizes the shapes of numbers? In order to learn, this type of artificial neural network needs to be provided with the correct answers. For each input image, it needs to know which of the ten possible numbers it corresponds to. The network can correct itself only by calculating the difference between its response and the correct answer. This procedure is known as "supervised learning": a supervisor, outside the system, knows the solution and tries to teach it to the machine. This is effective, but it should be noted that this situation, where the right answer is known in advance, is rather rare. When children learn to walk, no one tells them exactly which muscles to contract—they are simply encouraged again and again until they no longer fall. Babies learn solely on the basis of an evaluation of the result: I fell, or, on the contrary, I finally managed to walk across the room.

Artificial intelligence faces the same "unsupervised learning" problem. When a machine learns to play a video game, for example, the only thing it is told is that it must try to attain the highest score. No one tells it in advance what specific actions need to be taken to achieve this. How can it quickly find out for itself the right way of going about it?

Scientists have responded to this challenge by inventing "reinforcement learning," whereby we do not provide the system with any detail about what it must do (nobody knows!), but only with a "reward," an evaluation in the form of a quantitative score.[5] Even worse, the machine may receive its score after a delay, long after the decisive actions that led to it. Such delayed reinforcement learning is the principle by which the company DeepMind, a Google subsidiary, created a machine capable of playing chess, checkers, and Go. The problem is colossal for a simple reason: it is only at the very end that the system receives a single reward signal, indicating whether the game was won or lost. During the game itself, the system receives no feedback whatsoever—only the final checkmate counts. How, then, can the system figure out what to do at any given time? And, once the final score is known, how can the machine retrospectively evaluate its decisions?

The trick that computer scientists have found is to program the machine to do two things at the same time: to act and to self-evaluate. One half of the system, called the "critic," learns to predict the final score. The goal of this network of artificial neurons is to evaluate, as accurately as possible, the state of the game, in order to predict the final reward: Am I winning or losing? Is my balance stable, or am I about to fall? Thanks to this critic that emerges in this half of the machine, the system can evaluate its actions at every moment and not just at the end. The other half of the machine, the actor, can then use this evaluation to correct itself: Wait! I'd better avoid this or that action, because the critic thinks it will increase my chances of losing.

Trial after trial, the actor and the critic progress together: one learns to act wisely, focusing on the most effective actions, while the other learns to evaluate, ever more sharply, the consequences of these acts. In the end, unlike the famed guy who is falling from a skyscraper and exclaims, "So far, so good," the actor-critic network becomes endowed with a remarkable prescience: the ability to predict, within the vast seas of not-yet-lost games, those that are likely to be won and those that will lead only to disaster.

The actor-critic combination is one of the most effective strategies of contemporary artificial intelligence. When backed by a hierarchical neural network, it works wonders. As early as the 1980s, it enabled a neural network to win the backgammon world cup. More recently, it enabled DeepMind to create a multifunctional neural network capable of learning to play all kinds of video games such as *Super Mario* and *Tetris*.[6] One simply gives this system the pixels of the image as an input, the possible actions as an output, and the score of the game as a reward function. The machine learns everything else. When it plays *Tetris*, it discovers that the screen is made up of shapes, that the falling one is more important than the others, that various actions can change its orientation and its position, and so on and so forth—until the machine turns into an artificial player of formidable effectiveness. And when it plays *Super Mario*, the change in inputs and rewards teaches it to attend to completely different settings: what pixels form Mario's body, how he moves, where the enemies are, the shapes of walls, doors, traps, bonuses . . . and how to act in front of each of them. By adjusting its parameters, i.e., the millions of connections that link the layers together, a single network can adapt to all kinds of games and learn to recognize the shapes of *Tetris*, *Pac-Man*, or *Sonic the Hedgehog*.

What is the point of teaching a machine to play video games? Two years later, DeepMind engineers used what they had learned from game playing to solve an economic problem of vital interest: How should Google optimize the management of its computer servers? The artificial neural network remained similar; the only things that changed were the inputs (date, time, weather, international events, search requests, number of people connected to each server, etc.), the outputs (turn on or off this or that server on various continents), and the reward function (consume less energy). The result was an instant drop in power consumption. Google reduced its energy bill by up to 40 percent and saved tens of millions of dollars—even after myriad specialized engineers had already tried to optimize those very servers. Artificial intelligence has truly reached levels of success that can turn whole industries upside down.

DeepMind has achieved even more amazing feats. As everyone probably knows, its AlphaGo program managed to beat eighteen-time world champion Lee Sedol in the game of Go, considered until very recently the Everest of artificial intelligence.[7] This game is played on a vast square checkerboard (a *goban*) with nineteen positions on each side, for a total of 361 places where black and white pieces can be played. The number of combinations is so vast that it is strictly impossible to systematically explore all the future moves available to each player. And yet reinforcement learning allowed the AlphaGo software to recognize favorable and unfavorable combinations better than any human player. One of the many tricks was to make the system play against itself, just as a chess player trains by playing both white and black. The idea is simple: at the end of each game, the winning software strengthens its actions, while the loser weakens them—but both have also learned to evaluate their moves more efficiently.

We happily mock Baron Munchausen, who, in his fabled *Adventures*, foolishly attempts to fly away by pulling on his bootstraps. In artificial intelligence, however, Munchausen's mad method gave birth to a rather sophisticated strategy, aptly called "bootstrapping"—little by little, starting from a meaningless architecture devoid of knowledge, a neural network can become a world champion, simply by playing against itself.

This idea of increasing the speed of learning by letting two networks collaborate—or, on the contrary, compete—continues to lead to major advances in artificial intelligence. One of the most recent ideas, called "adversarial learning,"[8] consists of training two opponent systems: one that learns to become an expert (say, in Van Gogh's paintings) and another whose sole goal is to make the first one fail (by learning to become a brilliant forger of false Van Goghs). The first system gets a bonus whenever it successfully identifies a genuine Van Gogh painting, while the second is rewarded whenever it manages to fool the other's expert eye. This adversarial learning algorithm yields not just one but two artificial intelligences: a world authority in Van Gogh, fond of the smallest details that can authenticate a true painting by the master, and a genius forger, capable of producing paintings that can fool the best of experts. This sort of training can be likened to the preparation for a presidential debate: a candidate can sharpen her training by hiring someone to imitate her opponent's best lines.

Could this approach apply to a single human brain? Our two hemispheres and numerous subcortical nuclei also host a whole collection of experts who fight, coordinate, and evaluate one another. Some of the areas in our brain learn to simulate what others are doing; they allow us to foresee and imagine the results of our actions, sometimes with a realism worthy of the best counterfeiters: our memory and imagination can make us see the seaside bay where we swam last summer, or the door handle that we grab in the dark. Some areas learn to criticize others: they constantly assess our abilities and predict the rewards or punishments we might get. These are the areas that push us to act or to remain silent. We will also see that metacognition—the ability to know oneself, to self-evaluate, to mentally simulate what would happen if we acted this way or that way—plays a fundamental role in human learning. The opinions we form of ourselves help us progress or, in some cases, lock us into a vicious circle of failure. Thus, it is not inappropriate to think of the brain as a collection of experts that collaborate and compete.

LEARNING IS RESTRICTING SEARCH SPACE

Contemporary artificial intelligence still faces a major problem. The more parameters the internal model has, the more difficult it is to find the best way to adjust it. And in current neural networks, the search space is immense. Computer scientists therefore have to deal with a massive combinatorial explosion: at each stage, millions of choices are available, and their combinations are so vast that it is impossible to explore them all. As a result, learning is sometimes exceedingly slow: it takes billions of attempts to move the system in the right direction within this immense landscape of possibilities. And the data, however large, become scarce relative to the gigantic size of that space. This issue is called the "curse of dimensionality"—learning can become very hard when you have millions of potential levers to pull.

The immense number of parameters that neural networks possess often leads to a second obstacle, which is called "overfitting" or "overlearning": the system has so many degrees of freedom that it finds it easier to memorize all the details of each example than it is to identify a more general rule that can explain them.

As John von Neumann (1903–57), the father of computer science, famously said, "With four parameters I can fit an elephant, and with five I can make him wiggle his trunk." What he meant is that having too many free parameters can be a curse: it's all too easy to "overfit" any data simply by memorizing every detail, but that does not mean that the resulting system captures anything significant. You can fit the pachyderm's profile without understanding anything deep about elephants as a species. Having too many free parameters can be detrimental to abstraction. While the system easily learns, it is unable to generalize to new situations. Yet this ability to generalize is the key to learning. What would be the point of a machine that could recognize a picture that it has already seen, or win a game of Go that it has already played? Obviously, the real aim is to recognize any picture, or to win against any player, whether the circumstances are familiar or new.

Again, computer scientists are investigating various solutions to these problems. One of the most effective interventions, which can both accelerate learning and improve generalization, is to simplify the model. When the number of parameters to be adjusted is minimized, the system can be forced to find a more general solution. This is the key insight that led LeCun to invent *convolutional neural networks*, an artificial learning device which has become ubiquitous in the field of image recognition.[9] The idea is simple: in order to recognize the items in a picture, you pretty much have to do the same job everywhere. In a photo, for example, faces may appear anywhere. To recognize them, one should apply the same algorithm to every part of the picture (e.g., to look for an oval, a pair of eyes, etc.). There is no need to learn a different model at each point of the retina: what is learned in one place can be reused everywhere else.

Over the course of learning, LeCun's convolutional neural networks apply whatever they learn from a given region to the entire network, at all levels and on ever wider scales. They therefore have a much smaller number of parameters to learn: by and large, the system has to tune only a single filter that it applies everywhere, rather than a plethora of different connections for each location in the image. This simple trick massively improves performance, especially generalization to new images. The reason is simple: the algorithm that runs on a new image benefits from the immense experience it gained from every point of every photo that it has ever seen. It also speeds up learning, since the machine explores only a subset of vision models. Prior to learning, it already knows something important about the world: that the same object can appear anywhere in the image.

This trick generalizes to many other domains. To recognize speech, for example, one must abstract away from the specifics of the speaker's voice. This is achieved by forcing a neural network to use the same connections in different frequency bands, whether the voice is low or high. Reducing the number of parameters that must be adjusted leads to greater speeds and better generalization to new voices: the advantage is twofold, and this is how your smartphone is able to respond to your voice.

LEARNING IS PROJECTING A PRIORI HYPOTHESES

Yann LeCun's strategy provides a good example of a much more general notion: the exploitation of innate knowledge. Convolutional neural networks learn better and faster than other types of neural networks because they do not learn everything. They incorporate, in their very architecture, a strong hypothesis: what I learn in one place can be generalized everywhere else.

The main problem with image recognition is invariance: I have to recognize an object, whatever its position and size, even if it moves to the right or left, farther or closer. It is a challenge, but it is also a very strong constraint: I can expect the very same clues to help me recognize a face anywhere in space. By replicating the same algorithm everywhere, convolutional networks effectively exploit this constraint: they integrate it into their very structure. Innately, prior to any learning, the system already "knows" this key property of the visual world. It does not learn invariance, but assumes it a priori and uses it to reduce the learning space—clever indeed!

The moral here is that nature and nurture should not be opposed. Pure learning, in the absence of any innate constraints, simply does not exist. Any learning algorithm contains, in one way or another, a set of assumptions about the domain to be learned. Rather than trying to learn everything from scratch, it is much more effective to rely on prior assumptions that clearly delineate the basic laws of the domain that must be explored, and integrate these laws into the very architecture of the system. The more innate assumptions there are, the faster learning is (provided, of course, that these assumptions are correct!). This is universally true. It would be wrong, for example, to think that the AlphaGo Zero software, which trained itself in Go by playing against itself, started from nothing: its initial representation included, among other things, knowledge of the topography and symmetries of the game, which divided the search space by a factor of eight.

Our brain too is molded with assumptions of all kinds. Shortly, we will see that, at birth, babies' brains are already organized and knowledgeable. They know, implicitly, that the world is made of things that move only when pushed, without ever interpenetrating each other (solid objects)—and also that it contains much stranger entities that speak and move by themselves (people). No need to learn these laws: since they are true everywhere humans live, our genome hardwires them into the brain, thus constraining and speeding up learning. Babies do not have to learn everything about the world: their brains are full of innate constraints, and only the specific parameters that vary unpredictably (such as face shape, eye color, tone of voice, and individual tastes of the people around them) remain to be acquired.

Again, nature and nurture need not be opposed. If the baby's brain knows the difference between people and inanimate objects, it is because, in a sense, it has learned this—not in the first few days of its life, but in the course of millions of years of evolution. Darwinian selection is, in effect, a learning algorithm—an incredibly powerful program that has been running for hundreds of millions of years, in parallel, across billions of learning machines (every creature that ever lived).[10] We are the heirs of an unfathomable wisdom. Through Darwinian trial and error, our genome has internalized the knowledge of the generations that have preceded us. This innate knowledge is of a different type than the specific facts that we learn during our lifetime: it is much more abstract, because it biases our neural networks to respect the fundamental laws of nature.

In brief, during pregnancy, our genes lay down a brain architecture that guides and accelerates subsequent learning by imposing restrictions on the size of the explored space. In computer-science lingo, one may say that genes set up the "hyperparameters" of the brain: the high-level variables that specify the number of layers, the types of neurons, the general shape of their inter-connections, whether they are duplicated at any point on the retina, and so on and so forth. Because many of these variables are stored in our genome, we no longer need to learn them: our species internalized them as it evolved.

Our brain is therefore not simply passively subjected to sensory inputs. From the get-go, it already possesses a set of abstract hypotheses, an accumu-lated wisdom that emerged through the sift of Darwinian evolution and which it now projects onto the outside world. Not all scientists agree with this idea, but I consider it a central point: the naive empiricist philosophy underlying many of today's artificial neural networks is wrong. It is simply not true that we are born with completely disorganized circuits devoid of any knowledge, which later receive the imprint of their environment. Learning, in man and machine, always starts from a set of a priori hypotheses, which are projected onto the incoming data, and from which the system selects those that are best suited to the current environment. As Jean-Pierre Changeux stated in his best-selling book *Neuronal Man* (1985), "To learn is to eliminate."

CHAPTER 2

Why Our Brain Learns Better
Than Current Machines

THE RECENT SURGE OF PROGRESS IN ARTIFICIAL INTELLIGENCE MAY SUG-
gest that we have finally discovered how to copy and even surpass human learning and intelligence. According to some self-proclaimed prophets, machines are
about to overtake us. Nothing could be further from the truth. In fact, most
cognitive scientists, while admiring recent advances in artificial neural networks, are well aware of the fact that these machines remain highly limited.
In truth, most artificial neural networks implement only the operations that
our brain performs unconsciously, in a few tenths of a second, when it perceives an image, recognizes it, categorizes it, and accesses its meaning.[1] However, our brain goes much further: it is able to explore the image consciously,
carefully, step by step, for several seconds. It formulates symbolic representations and explicit theories of the world that we can share with others through
language.

Operations of this nature—slow, reasoned, symbolic—remain (for now) the
exclusive privilege of our species. Current machine learning algorithms capture them poorly. Although there is constant progress in the fields of machine
translation and logical reasoning, a common criticism of artificial neural networks is that they attempt to learn everything at the same level, as if every
problem were a matter of automatic classification. To a man with a hammer,

everything looks like a nail! But our brain is much more flexible. It quickly manages to prioritize information and, whenever possible, extract general, logical, and explicit principles.

WHAT IS ARTIFICIAL INTELLIGENCE MISSING?

It is interesting to try to clarify what artificial intelligence is still missing, because this is also a way to identify what is unique about our species' learning abilities. Here is a short and probably still partial list of functions that even a baby possesses and that most current artificial systems are missing:

Learning abstract concepts. Most artificial neural networks capture only the very first stages of information processing—those that, in less than a fifth of a second, parse an image in the visual areas of our brain. Deep learning algorithms are far from being as deep as some people claim. According to Yoshua Bengio, one of the inventors of deep learning algorithms, they actually tend to learn superficial statistical regularities in data, rather than high-level abstract concepts.[2] To recognize an object, for instance, they often rely on the presence of a few shallow features in the image, such as a specific color or shape. Change these details and their performance collapses: contemporary convolutional neural networks are unable to recognize what constitutes the essence of an object; they have difficulty understanding that a chair remains a chair whether it has four legs or just one, and whether it is made of glass, metal, or inflatable plastic. This inclination to attend to superficial features makes these networks susceptible to massive errors. There is a whole literature on how to fool a neural network: take a banana and modify a few pixels or put a particular sticker on it, and the neural network will think it's a toaster!

True enough, when you flash an image to a person for a split second, they will sometimes make the same kinds of errors as a machine and may mistake a dog for a cat.[3] However, as soon as humans are given a little more time, they correct their errors. Unlike a computer, we possess the ability to question our beliefs and refocus our attention on those aspects of an image that do not fit with our first impression. This second analysis, conscious and intelligent, calls upon our general powers of reasoning and abstraction. Artificial neural networks neglect an essential point: human learning is not just the setting of a pattern-recognition filter, but the forming of an abstract model of the world. By learning to read, for example, we have acquired an abstract concept of each letter of the alphabet, which allows us to recognize it in all its disguises, as well as generate new versions:

A A A A 𝒜 A Λ A 𝒜

The cognitive scientist Douglas Hofstadter once said that the real challenge for artificial intelligence was to recognize the letter A! This quip was undoubtedly an exaggeration, but a profound one nevertheless: even in this most trivial context, humans deploy an unmatched knack for abstraction. This feat is at the origin of an amusing occurrence of daily life: the CAPT-CHA, the little chain of letters that some websites ask you to recognize in order to prove you are a human being, not a machine. For years, CAPT-CHAs have withstood machines. But computer science is evolving fast: in 2017, an artificial system managed to recognize CAPTCHAs at an almost humanlike level.[4] Unsurprisingly, this algorithm mimics the human brain in several respects. A genuine tour de force, it manages to extract the skeleton of each letter, the inner essence of the letter A, and uses all the resources of statistical reasoning to verify whether this abstract idea applies to the current image. Yet this computer algorithm, however sophisticated, applies only to CAPTCHAs. Our brains apply this ability for abstraction to all aspects of our daily lives.

Data-efficient learning. Everyone agrees that today's neural networks learn far too slowly: they need thousands, millions, even billions of data points to develop an intuition of a domain. We even have experimental evidence of this sluggishness. For instance, it takes no less than nine hundred hours of play for the neural network designed by DeepMind to reach a reasonable level on an Atari console—while a human being reaches the same level in two hours![5] Another example is language learning. Psycholinguist Emmanuel Dupoux estimates that in most French families, children hear about five hundred to one thousand hours of speech per year, which is more than enough for them to acquire Descartes's patois, including such quirks as *soixante-douze* or *s'il vous plaît*. However, among the Tsimane, an indigenous population of the Bolivian Amazon, children hear only sixty hours of speech per year—and remarkably, this limited experience does not prevent them from becoming excellent speakers of the Tsimane language. In comparison, the best current computer systems from Apple, Baidu, and Google require anywhere between twenty and a thousand times more data in order to attain a modicum of language competence. In the field of learning, the effectiveness of the human brain remains unmatched: machines are data hungry, but humans are data efficient. Learning, in our species, makes the most from the least amount of data.

Social learning. Our species is the only one that voluntarily shares information: we learn a lot from our fellow humans through language. This ability remains beyond the reach of current neural networks. In these models, knowledge is encrypted, diluted in the values of hundreds of millions of synaptic weights. In this hidden, implicit form, it cannot be extracted and selectively shared with others. In our brains, by contrast, the highest-level information, which reaches our consciousness, can be explicitly stated to others. Conscious knowledge comes with verbal reportability: whenever we understand something in a sufficiently perspicuous manner, a mental formula resonates in our language of thought, and we can use the words of language to report it. The extraordinary efficiency with which we manage to share our knowledge with others, using a minimum number of words ("To get to the market, turn right on the small street behind the church."), remains unequalled, in the animal kingdom as in the computer world.

One-trial learning. An extreme case of this efficiency is when we learn something new on a single trial. If I introduce a new verb, let's say *purget*, even only once, it will be enough for you to use it. Of course, some artificial neural networks are also capable of storing a specific episode. But what machines cannot yet do well, and that the human brain succeeds in doing wonderfully, is integrate new information within an existing network of knowledge. You not only memorize the new verb *purget*, but you immediately know how to conjugate it and insert it into other sentences: Do you ever purget? I purgot it yesterday. Have you ever purgotten? Purgetting is a problem. When I say, "Let's purget tomorrow," you don't just learn a word—you also insert it into a vast system of symbols and rules: it is a verb with irregular past tense (purgot, purgotten) and a typical conjugation in the present tense (I purget, you purget, she purgets, etc.). To learn is to succeed in inserting new knowledge into an existing network.

Systematicity and the language of thought. Grammar rules are just one example of a particular talent in our brain: the ability to discover the general laws that lie behind specific cases. Whether it is in mathematics, language, science, or music, the human brain manages to extract very abstract principles, systematic rules that it can reapply in many different contexts. Take arithmetic, for example: our ability to add two numbers is extremely general—once we have learned this procedure with small numbers, we can systematize it to arbitrarily large numbers. Better yet, we can draw inferences of extraordinary generality. Many children, around five or six years of age, discover that each number *n* has a successor *n + 1*, and that the sequence of whole numbers is therefore infinite—there is no greatest number. I still remember, with emotion, the moment when I became aware of this—it was, in reality, my first mathematical theorem. What extraordinary powers of abstraction! How does our brain, which includes a finite number of neurons, manage to conceptualize infinity?

Present-day artificial neural networks cannot represent an abstract law as simple as "every number has a successor." Absolute truths are not their cup of tea. Systematicity,[6] the ability to generalize on the basis of a symbolic rule rather than a superficial resemblance, still eludes most current algorithms. Ironically, so-called deep learning algorithms are almost entirely incapable of any profound insight.

Our brain, on the other hand, seems to have a flowing ability to conceive formulas in a kind of mental language. For instance, it can express the concept of an infinite set because it possesses an internal language endowed with such abstract functions as negation and quantification (infinite = *not* finite = beyond *any* number). The American philosopher Jerry Fodor (1935–2017) theorized this ability: he postulated that our thinking consists of symbols that combine according to the systematic rules of a "language of thought."[7] Such a language owes its power to its recursive nature: each newly created object (say, the concept of infinity) can immediately be reused in new combinations, without limits. How many infinities are there? Such is the seemingly absurd question that the mathematician Georg Cantor (1845–1918) asked himself, which led him to formulate the theory of transfinite numbers. The ability to "make infinite use of finite means," according to Wilhelm von Humboldt (1767–1835), characterizes human thought.

Some computer-science models try to capture the acquisition of abstract mathematical rules in children—but to do so, they have to incorporate a very different form of learning, one that involves rules and grammars and manages to quickly select the shortest and most plausible of them.[8] In this view, learning becomes similar to programming: it consists of selecting the simplest internal formula that fits the data, among all those available in the language of thought.

Current neural networks are largely unable to represent the range of abstract phrases, formulas, rules, and theories with which the *Homo sapiens* brain models the world. This is probably no coincidence: there is something profoundly human about this, something that is not found in the brains of other animal species, and that contemporary neuroscience has not yet managed to address—a genuinely singular aspect of our species. Among primates, our brain seems to be the only one to represent sets of symbols that combine according to a complex and arborescent syntax.[9] My laboratory, for example, has shown that the human brain cannot help hearing a series of sounds such as *beep beep beep boop* without immediately theorizing the underlying abstract structure (three identical sounds followed by a different one). Placed in the same situation, a monkey detects a series of four sounds, realizes that the last is different, but does not seem to integrate this piecewise knowledge into a single formula; we know this because when we examine their brain activity, we see distinct circuits activate for number and for sequence, but never observe the integrated pattern of activity that we find in the human language area called "Broca's area."[10]

Similarly, it takes tens of thousands of trials before a monkey understands how to reverse the order of a sequence (from ABCD to DCBA), while for a four-year-old human, five trials are enough.[11] Even a baby of a few months of age already encodes the external world using abstract and systematic rules—an ability that completely eludes both conventional artificial neural networks and other primate species.

Composition. Once I have learned, say, to add two numbers, this skill becomes an integral part of my repertoire of talents: it becomes immediately available to address all my other goals. I can use it as a subroutine in dozens of different contexts, for example, to pay the restaurant bill or to check my tax forms. Above all, I can recombine it with other learned skills—I have no difficulty, for example, following an algorithm that asks me to take a number, add two, and decide whether it is now larger or smaller than five.[12]

It is surprising that current neural networks do not yet show this flexibility. The knowledge that they have learned remains confined in hidden, inaccessible connections, thus making it very difficult to reuse in other, more complex tasks. The ability to *compose* previously learned skills, that is, to recombine them in order to solve new problems, is beyond these models. Today's artificial intelligence solves only extremely narrow problems: the AlphaGo software, which can defeat any human champion in Go, is a stubborn expert, unable to generalize its talents to any other game even slightly different (including the game of Go on a fifteen-by-fifteen board rather than the standard nineteen-by-nineteen goban). In the human brain, on the other hand, learning almost always means rendering knowledge explicit, so that it can be reused, recombined, and explained to others. Here again, we are dealing with a singular aspect of the human brain, linked to language and which has proven to be difficult to reproduce in a machine. As early as 1637, in his famous *Discourse on Method*, Descartes anticipated this issue:

> If there were machines that resembled our bodies and imitated our actions as much as is morally possible, we would always have two very certain means for recognizing that they are not genuinely human. The first is that they would never be able to use speech, or other signs by composing them, as we do to express our thoughts to others. For one could easily conceive of a machine that is made in such a way that it utters words . . . but it could not arrange words in different ways to reply to the meaning of everything that is said in its presence, as even the most unintelligent human beings can do. And the second means is that, even if they did many things as well as or, possibly, better than anyone of us, they would infallibly fail in others. One would thus discover that they did not act on the basis of knowledge, but merely as a result of the disposition of their organs. For whereas reason is a universal instrument that can be used in all kinds of situations, these organs need a specific disposition for every particular action.

Reason, the mind's universal instrument. . . . The mental abilities that Descartes lists point to a second learning system, hierarchically higher than the previous one, based on rules and symbols. In its early stages, our visual system vaguely resembles current artificial neural networks: it learns to filter incoming images and to recognize frequent configurations. This suffices to recognize a face, a word, or a configuration of the game Go. But then the processing style radically changes: learning begins to resemble reasoning, a logical inference that attempts to capture the rules of a domain. Creating machines that reach this second level of intelligence is a great challenge for contemporary artificial intelligence research. Let's examine two elements that define what humans do when they learn at this second level, and that defy most current machine learning algorithms.

LEARNING IS INFERRING THE GRAMMAR OF A DOMAIN

Characteristic of the human species is a relentless search for abstract rules, high-level conclusions that are extracted from a specific situation and subsequently tested on new observations. Attempting to formulate such abstract laws can be an extraordinarily powerful learning strategy, since the most abstract laws are precisely those that apply to the greatest number of observations. Finding the appropriate law or logical rule that accounts for all available data is the ultimate means to massively accelerate learning—and the human brain is exceedingly good at this game.

Let us consider an example. Imagine that I show you a dozen opaque boxes filled with balls of different colors. I select a box at random, one from which I have never drawn anything out before. I plunge my hand into it, and I draw a green ball. Can you deduce anything about the contents of the box? What color will the next ball be?

The first answer that probably comes to mind is: I have no idea—you gave me practically no information; how could I know the color of the next ball? Yes, but . . . imagine that, in the past, I drew some balls from the other boxes and you noticed the following rule: in a given box, all balls are always the same color. The problem becomes trivial. When I show you a new box, you need only to draw a single green ball to deduce that all the other balls will be this color. With this general rule in mind, it becomes possible to learn in a single trial.

This example illustrates how higher-order knowledge, formulated at what is often called the "meta" level, can guide a whole set of lower-level observations. The abstract meta-rule that "in a given box, all the balls are the same color," once learned, massively accelerates learning. Of course, it may also turn out to be false. You will then be massively surprised (or should I say "meta-surprised") if the tenth box you explore contains balls of all colors. In this case, you would have to revise your mental model and question the assumption that all boxes are similar. Perhaps you would propose an even higher-level hypothesis, a meta-meta-hypothesis—for instance, you may suppose that boxes come in two kinds, single-colored and multicolored, in which case you would need at least two draws per box before concluding anything. In any case, formulating a hierarchy of abstract rules would save you valuable learning time.

Learning, in this sense, therefore means managing an internal hierarchy of rules and trying to infer, as soon as possible, the most general ones that summarize a whole series of observations. The human brain seems to apply this hierarchical principle from childhood. Take a two- or three-year-old child walking in a garden and learning a new word from his or her parents, say, *butterfly*. Often, it is enough for the child to hear the word once or twice, and voilà: its meaning is memorized. Such a learning speed is amazing. It surpasses every known artificial intelligence system to date. Why is the problem difficult? Because every utterance of every word does not fully constrain its meaning. The word *butterfly* is typically uttered as the child is immersed in a complex scene, full of flowers, trees, toys, and people; all of these are potential candidates for the meaning of that word—not to mention the less obvious meanings: every moment we live is full of sounds, smells, movements, actions, but also abstract properties. For all we know, butterfly could mean color, sky, move, or symmetry. The existence of abstract words makes this problem most perplexing. How do children learn the meanings of the words *think*, *believe*, *no*, *freedom*, and *death*, if the referents cannot be perceived or experienced? How do they understand what "I" means, when each time they hear it, the speakers are talking about . . . themselves?!

The fast learning of abstract words is as incompatible with naive views of word learning as Pavlovian conditioning or Skinnerian association. Neural networks that simply try to correlate inputs with outputs and images with words, typically require thousands of trials before they begin to understand that the word *butterfly* refers to that colored insect, there, in the corner of the image . . . and such a shallow correlation of words with pictures will never discover the meanings of words without a fixed reference, such as *we*, *always*, or *smell*.

Word acquisition poses a huge challenge to cognitive science. However, we know that part of the solution lies in the child's ability to formulate non-linguistic, abstract, logical representations. Even before they acquire their first words, children possess a kind of language of thought within which they can formulate and test abstract hypotheses. Their brains are not blank slates, and the innate knowledge that they project onto the external world can drastically restrict the abstract space within which they learn. Furthermore, children quickly learn the meaning of words because they select among hypotheses using as a guide a whole panoply of high-level rules. Such meta-rules massively accelerate learning, exactly as in the problem of the colored balls in the different boxes.

One of these rules that facilitates vocabulary acquisition is to always favor the simplest, smallest assumption compatible with the data. For instance, when a baby hears its mother say, "Look at the dog," in theory, nothing precludes the word *dog* from referring to *that* particular dog (Snoopy)—or, conversely, any mammal, four-legged creature, animal, or living being. How do children discover the true meaning of a word—that *dog* means all dogs, but only dogs? Experiments suggest that they reason logically by testing all hypotheses but keeping only the simplest one that fits with what they heard. Thus, when children hear the word *Snoopy*, they always hear it in the context of that specific pet, and the smallest set compatible with those observables is confined to that particular dog. And the first time children hear the word *dog*, in a single specific context, they may temporarily believe that the word refers to only that particular animal—but as soon as they hear it twice, in two different contexts, they can infer that the word refers to a whole category. A mathematical model of this process predicts that three or four instances are enough to converge toward the appropriate meaning.[13] This is the inference that children make, faster than any current artificial neural network.

Other tricks allow children to learn language in record time, compared with present-day AI systems. One of these meta-rules expresses a truism: in general, the speaker pays attention to what he or she is talking about. Once babies understand this rule, they can considerably restrict the abstract space in which they search for meaning: they do not have to correlate every word with all the objects present in the visual scene, as a computer would, until they obtain enough data to prove that each time they hear about a butterfly, the little colored insect is present. All the child has to do to infer what his mother is talking about is follow her gaze or the direction of her finger: this is called "shared attention," and it is a fundamental principle of language learning.

Here is an elegant experiment: Take a two- or three-year-old child, show him a new toy, and have an adult look at it while saying, "Oh, a wog!" A single trial suffices for the child to figure out that *wog* is the name of that object. Now replicate the situation, except that the adult doesn't say a word, but the child hears "Oh, a wog!" uttered by a loudspeaker on the ceiling. The child learns strictly nothing, because he can no longer decipher the speaker's intention.[14] Babies learn the meaning of a new word only if they manage to understand the intention of the person who uttered it. This ability also enables them to acquire a lexicon of abstract words: to do so, they must put themselves in the speaker's place to understand which thought or word the speaker intended to refer to.

Children use many other meta-rules to learn words. For example, they capitalize on grammatical context: when they are told, "Look at the butterfly," the presence of the determiner word *the* makes it very likely that the following word is a noun. This is a meta-rule that they had to learn—babies obviously are not born with an innate knowledge of all possible articles in every language. However, research shows that this type of learning is fast: by twelve months, children have already recorded the most frequent determiners and other function words and use them to guide subsequent learning.[15]

They are able to do this because these grammatical words are very frequent and, whenever they appear, almost invariably precede a noun or a noun phrase. The reasoning may seem circular, but it is not: babies start learning their first nouns, beginning with extremely familiar ones like *bottle* and *chair*, around six months of age . . . then they notice that these words are often preceded by a very frequent word, the article *the* . . . from which they deduce that all these words probably belong to the same category, *noun* . . .

Here are three tufas. Can you identify the others?

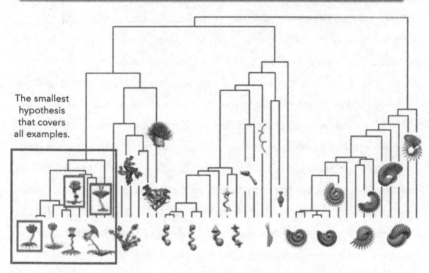

Learning means trying to select the simplest model that fits the data. Suppose I show you the top card and tell you that the three objects surrounded by thick lines are "tufas." With so little data, how do you find the other tufas? Your brain makes a model of how these forms were generated, a hierarchical tree of their properties, and then selects the smallest branch of the tree which is compatible with all the data.

and that these words often refer to things ... a meta-rule which enables them, when they hear a new utterance, such as "the butterfly," to first search for a possible meaning among the objects around them, rather than treating the word as a verb or an adjective. Thus, each learning episode reinforces this rule, which itself facilitates subsequent learning, in a vast movement that accelerates every day. Developmental psychologists say that the child relies on syntactic bootstrapping: a children's language-learning algorithm manages to take off gradually, on its own, by capitalizing on a series of small but systematic inference steps.

There is yet another meta-rule that children use to speed up word learning. It is called the "mutual exclusivity assumption" and can be stated succinctly: one name for each thing. The law basically says that it is unlikely for two different words to refer to the same concept. A new word therefore most probably refers to a new object or idea. With this rule in mind, as soon as children hear an unfamiliar word, they can restrict their search for meaning to things whose names they do not yet know. And, as of sixteen months of age, children use this trick quite astutely.[16] Try the following experiment: take two bowls, a blue one and another of an unusual color—say, olive green—and tell the child, "Give me the tawdy bowl." The child will give you the bowl that is not blue (a word he already knows)—he seems to assume that if you had wanted to speak about the blue bowl, you would have used the word *blue*; ergo, you must be referring to the other, unknown one. Weeks later, that single experience will suffice for him to remember that this odd color is "tawdy."

Here again, we see how the mastery of a meta-rule can massively accelerate learning. And it is likely that this meta-rule itself was learned. Indeed, some experiments indicate that children from bilingual families apply this rule much less than monolingual babies.[17] Their bilingual experience makes them realize that their parents *can* use different words to say the same thing. Monolingual children, on the other hand, heavily rely on the exclusivity rule. They have figured out that whenever you use a new word, it is likely that you wanted them to learn a novel object or concept. If you say, "Give me the glax" in a room full of familiar objects, they will search everywhere for this mysterious object to which you are referring—and won't imagine that you could be referring to one of the known ones.

All these meta-rules illustrate what is called the "blessing of abstraction":

the most abstract meta-rules can be the easiest things to learn, because every word that the child hears provides evidence for them. Thus, the grammatical rule "nouns tend to be preceded by the article *the*" may well be acquired early on and guide the subsequent acquisition of a vast repertoire of nouns. Thanks to the blessing of abstraction, around two to three years of age, children enter a blessed period rightfully called the "lexical explosion," during which they effortlessly learn between ten and twenty new words a day, solely based on tenuous clues that still stall the best algorithms on the planet.

The ability to use meta-rules seems to require a good dose of intelligence. Does that make it unique to the human species? Not entirely. To some degree, other animals are also capable of abstract inference. Take the case of Rico, a shepherd dog who was trained to fetch a diverse range of objects.[18] All you have to do is say: "Rico, go fetch the dinosaur" . . . and the animal goes into the game room and comes back a few seconds later with a stuffed dinosaur in his mouth. The ethologists who tested him showed that Rico knows about two hundred words. But the most extraordinary thing is that he too uses the mutual exclusivity principle to learn new words. If you tell him, "Rico, go fetch the sikirid" (a new word), he always returns with a new object, one whose name he does not yet know. He too uses meta-rules such as "one name for each thing."

Mathematicians and computer scientists have begun to design algorithms that allow machines to learn such a hierarchy of rules, meta-rules, and meta-meta-rules, up to an arbitrary level. In these hierarchical learning algorithms, each learning episode constrains not only the low-level parameters, but also the knowledge of the highest level, the abstract hyperparameters which, in turn, will bias subsequent learning. While they do not yet imitate the extraordinary efficiency of language learning, these systems do achieve remarkable performance. For example, figure 4 in the color insert shows how a recent algorithm behaves as a kind of artificial scientist who finds the best model of the outside world.[19] This system possesses a set of abstract primitives, as well as a grammar that allows it to generate an infinity of higher-level structures through the recombination of these elementary rules. It can, for instance, define a linear chain as a set of closely connected points which is character-ized by the rule "each point has two neighbors, one to the left, one to the right"—and the system manages to discover, all by itself, that such a chain is the best way to represent the set of integers (a line that goes from zero to infinity) or politicians (from the ultraleft to the far right). A variant of the same grammar produces a binary tree where each node has one parent and two children. Such a tree structure is automatically selected when the system is asked to represent living beings—the machine, like an artificial Darwin, spontaneously rediscovers the tree of life!

Other combinations of rules generate planes, cylinders, and spheres, and the algorithm figures out how such structures approximate the geography of our planet. More sophisticated versions of the same algorithm manage to express even more abstract ideas. For example, American computer scientists Noah Goodman and Josh Tenenbaum designed a system capable of discover-ing the principle of causality[20]—the very idea that some events cause others. Its formulation is abstruse and mathematical: "In a directed, acyclic graph linking various variables, there exists a subset of variables on which all others depend." Although this expression is almost incomprehensible, I cite it because it nicely illustrates the kind of abstract internal formulas that this mental grammar is capable of expressing and testing. The system puts thousands of such formulas to the test and keeps only those that fit with the incoming data. As a result, it quickly infers the principle of causality (if, indeed, some of the sensory experiences it receives are causes and others are consequences). This is yet another illustration of the blessing of abstraction: entertaining such a high-level hypothesis massively accelerates learning, because it dramatically narrows the space of plausible hypotheses within which to search. And thanks to it, generations of children are on the lookout for explanations, constantly asking "Why?" and searching for causes—thus fueling our species's endless pursuit of scientific knowledge.

According to this view, learning consists of selecting, from a large set of expressions in the language of thought, the one that best fits the data. We will soon see that this is an excellent model of what children do. Like budding scientists, they formulate theories and compare them with the outside world. This implies that children's mental representations are much more structured than those of present-day artificial neural networks. From birth, the child's brain must already possess two key ingredients: all the machinery that makes it possible to generate a plethora of abstract formulas (a combinatorial language of thought) and the ability to choose from these formulas wisely, according to their plausibility given the data.

Such is the new vision of the brain:[21] an immense generative model, massively structured and capable of producing myriad hypothetical rules and structures—but which gradually restricts itself to those that fit with reality.

LEARNING IS REASONING AS A SCIENTIST

How does the brain select the best-fitting hypothesis? On what criteria should it accept or reject a model of the outside world? It turns out that there is an ideal strategy for doing so. This strategy lies at the very core of one of the most recent and productive theories of learning: the hypothesis that the brain behaves like a budding scientist. According to this theory, learning is reasoning like a good statistician who chooses, among several alternative theories, that which has the greatest probability of being correct, because it best accounts for the available data.

How does scientific reasoning work? When scientists formulate a theory, they do not just write down mathematical formulas—they make predictions. The strength of a theory is judged by the richness of the original predictions that emerged from it. The subsequent confirmation or rebuttal of those predictions is what leads to a theory's validation or downfall. Researchers apply a simple logic: they state several theories, unravel the web of ensuing predictions, and eliminate the theories whose predictions are invalidated by experiments and observations. Of course, a single experiment rarely suffices: it is often necessary to replicate the experiment several times, in different labs, in order to disentangle what is true from what is false. To paraphrase philosopher of science Karl Popper (1902–94), ignorance continuously recedes as a series of conjectures and refutations permit the progressive refinement of a theory.

The slow process of science resembles the way we learn. In each of our minds, ignorance is gradually erased as our brain successfully formulates increasingly accurate theories of the outside world through observations. But is this nothing more than a vague metaphor? No—it is, in fact, a rather precise statement about what the brain must be computing. And over the last thirty years, the hypothesis of "the child as a scientist" has led to a series of major discoveries about how children reason and learn.

Mathematicians and computer scientists have long theorized the best way of reasoning in the presence of uncertainties. This sophisticated theory is called "Bayesian," after its discoverer, the Reverend Thomas Bayes (1702–61), an English Presbyterian pastor and mathematician who became a member of the Royal Society. But perhaps we should be calling it the Laplacian theory, since it was the great French mathematician Pierre-Simon, Marquis de Laplace (1749–1827), who gave it its first complete formalization. In spite of its ancient roots, it is only in the last twenty years or so that this view has gained prominence in cognitive science and machine learning. An increasing number of researchers have begun to realize that only the Bayesian approach, firmly grounded in probability theory, guarantees the extraction of a maximum of information from each data point. To learn is to be able to draw as many inferences as possible from each observation, even the most uncertain ones—and this is precisely what Bayes's rule guarantees.

What did Bayes and Laplace discover? Simply put: the right way to make inferences, by reasoning with probabilities in order to trace every observation, however tenuous, back to its most plausible cause. Let's return to the foundations of logic. Since ancient times, humanity has understood how to reason with values of truth, *true* or *false*. Aristotle introduced the rules of deduction that we call syllogisms, which we all apply more or less intuitively. For example, the rule called "*modus tollens*" (literally translated as "method of denying") says that if P implies Q and it turns out that Q is false, then P must be false. It is this rule that Sherlock Holmes applied in the famous story "Silver Blaze":

"Is there any other point to which you would wish to draw my attention?" asks Inspector Gregory of Scotland Yard.

Holmes: "To the curious incident of the dog in the night-time."

Gregory: "The dog did nothing in the night-time."

Holmes: "That was the curious incident."

Sherlock reasoned that *if* the dog had spotted a stranger, *then* he would have barked. Because he did not, the criminal must have been a familiar person . . . reasoning that allows the famous detective to narrow down his search and eventually unmask the culprit.

"What does this have to do with learning?" you may be asking yourself. Well, learning is also reasoning like a detective: it always boils down to going back to the hidden causes of phenomena, in order to deduce the most plausible model that governs them. But in the real world, observations are rarely true or false: they are uncertain and probabilistic. And that is exactly where the fundamental contributions of the Reverend Bayes and the Marquis de Laplace come into play: Bayesian theory tells us how to reason with probabilities, what kinds of syllogisms we must apply when the data are not perfect, true or false, but probabilistic.

Probability Theory: The Logic of Science is the title of a fascinating book on Bayesian theory by statistician E. T. Jaynes (1922–98).[22] In it, he shows that what we call probability is nothing more than the expression of our uncertainty. The theory expresses, with mathematical precision, the laws according to which uncertainty must evolve when we make a new observation. It is the perfect extension of logic to the foggy domain of probabilities and uncertainties.

Let's take an example, similar in spirit to the one on which the Reverend Bayes founded his theory in the eighteenth century. Suppose I see someone flip a coin. If the coin is fair, it is equally likely to land on heads as it is tails: fifty-fifty. From this premise, classical probability theory tells us how to compute the chances of observing certain outcomes (for example, the probability of getting five tails in a row). Bayesian theory allows us to travel in the opposite direction, from observations to causes. It tells us, in a mathematically precise manner, how to answer such questions as "After several coin flips, should I change my views on the coin?" The default assumption is that the coin is unbiased . . . but if I see it land on tails twenty times, I have to revise my assumptions: the coin is most certainly rigged. Obviously, my original hypothesis has become improbable, but by how much? The theory precisely explains how to update our beliefs after each observation. Each assumption is assigned a number that corresponds to a plausibility or confidence level. With each observation, this number changes by a value that depends on the degree of improbability of the observed outcome. Just as in science, the more improbable an experimental observation is, the more it violates the predictions of our initial theory, and the more confidently we can reject that theory and look for alternative interpretations.

Bayesian theory is remarkably effective. During the Second World War, British mathematician Alan Turing (1912–54) used it to decrypt the Enigma code. At the time, German military messages were encrypted by the Enigma machine, a complex contraption of gears, rotors, and electrical cables, assembled to produce over a billion different configurations that changed after each letter. Every morning, the cryptographer would place the machine in the specific configuration that was planned for that day. He would then type a text, and Enigma would spit out a seemingly random sequence of letters, which only the owner of the encryption key could decode. To anyone else, the text seemed totally devoid of any order. But herein lies Turing's genius: he discovered that if two machines had been initialized in the same way, it introduced a slight bias in the distribution of letters, so that the two messages were slightly more likely to resemble each other. This bias was so small that no single letter was enough to conclude anything for certain. By accumulating those improbabilities, however, letter after letter, Turing could progressively gather more and more evidence that the same configuration had indeed been used twice. On this basis, and with the help of what they whimsically called "the bomb" (a large, ticking electro-mechanical machine that prefigured our computers), he and his team regularly broke the Enigma code.

Again, what's the relevance to our brains? Well, the very same type of reasoning seems to occur inside our cortex.[23] According to this theory, each region of the brain formulates one or more hypotheses and sends the corresponding predictions to other regions. In this way, each brain module constrains the assumptions of the next one, by exchanging messages that convey probabilistic predictions about the outside world. These signals are called "top-down" because they start in high-level cerebral areas, such as the frontal cortex, and make their way down to the lower sensory areas, such as the primary visual cortex. The theory proposes that these signals express the realm of hypotheses that our brain considers plausible and wishes to test.

In sensory areas, these top-down assumptions come into contact with "bottom-up" messages from the outside world, for instance, from the retina. At this moment, the model is confronted with reality. The theory says that the brain should calculate an error signal: the difference between what the model predicted and what has been observed. The Bayesian algorithm then indicates how to use this error signal to modify the internal model of the world. If there is no mistake, it means that the model was right. Otherwise, the error signal moves up the chain of brain areas and adjusts the model parameters along the way. Relatively quickly, the algorithm converges toward a mental model that fits the outside world.

According to this vision of the brain, our adult judgments combine two levels of insights: the innate knowledge of our species (what Bayesians call *priors*, the sets of plausible hypotheses inherited throughout evolution) and our personal experience (the *posterior*: the revision of those hypotheses, based on all the inferences we have been able to gather throughout our life). This division of labor puts the classic "nature versus nurture" debate to rest: our brain organization provides us with both a powerful start-up kit and an equally powerful learning machine. All knowledge must be based on these two components:

first, a set of a priori assumptions, prior to any interaction with the environment, and second, the capacity to sort them out according to their a posteriori plausibility, once we have encountered some real data.

One can mathematically demonstrate that the Bayesian approach is the best way to learn. This is the only way to extract the very essence of a learning episode and get the most out of it. Even a few bits of information, such as the suspicious coincidences that Turing spotted in the Enigma code, may suffice to learn. Once the system processes them, like a good statistician patiently accumulating evidence, it will inevitably end up with enough data to refute certain theories and validate others.

Is this really how the brain works? Is it capable of generating, at birth, vast realms of hypotheses that it learns to pick from? Does it proceed by elimination, selecting hypotheses according to how well the observed data support them? Do babies, right from birth, act as clever statisticians? Are they able to extract as much information as possible from each learning experience? Let's now take a closer look at the experimental data on babies' brains.

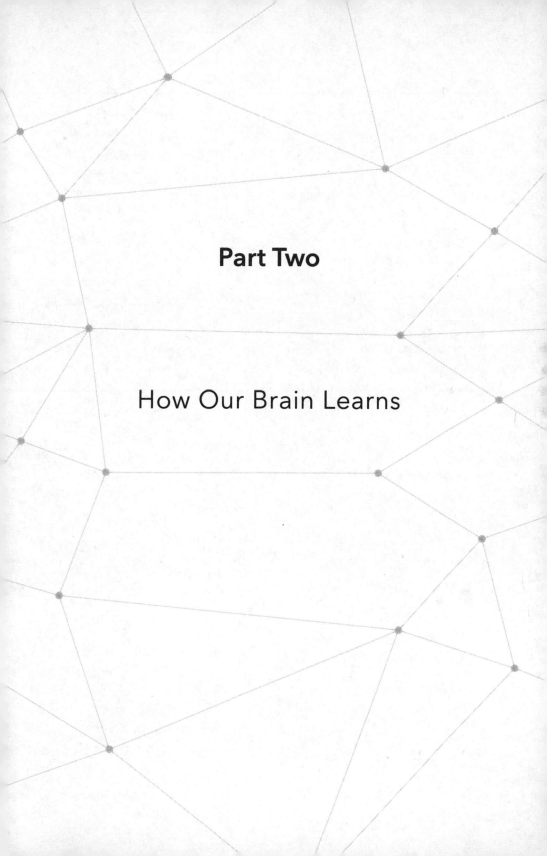

Part Two

How Our Brain Learns

The debate of nature versus nurture has raged for millennia. Are babies comparable to a white page, a blank slate, or an empty bottle that experience must fill? As early as 400 BCE, in *The Republic*, Plato was already rejecting the idea that our brains enter the world devoid of any knowledge. From birth, he claimed, every soul possesses two sophisticated mechanisms: the power of knowledge and the organ by which we acquire instruction.

As we have just seen, two thousand years later, a remarkably similar conclusion arose from advances in machine learning. Learning is vastly more effective when the machine comes equipped with two features: a vast space of hypotheses, a set of mental models with myriad settings to choose from; and sophisticated algorithms that adjust those settings according to the data received from the outside world. As one of my friends once said, in the debate on nature versus nurture, we have underestimated both! Learning requires two structures: an immense set of potential models and an efficient algorithm to adjust them to reality.

Artificial neural networks do this in their own way, by entrusting the representation of mental models to millions of adjustable connections. However, these systems, while capturing the rapid and unconscious recognition of images or speech, are not yet able to represent more abstract hypotheses, such as grammar rules or the logic of mathematical operations.

The human brain seems to function in a different way: our knowledge grows through the combination of symbols. According to this view, we come into the world with a vast number of possible combinations of potential thoughts. This language of thought, endowed with abstract assumptions and grammar rules, is already in place prior to learning. It generates a vast realm of hypotheses to be put to the test. And to do so, according to the Bayesian brain theory, our brain must act like a scientist, collecting statistical data and using them to select the best-fitting generative model.

This view of learning may seem counterintuitive. It suggests that each human baby's brain potentially contains all the languages of the world, all the objects, all the faces, and all the tools that it will ever encounter, in addition to all the words, all the facts, and all the events that it will ever remember. The combinatorics of the brain are such that all these objects of thought are potentially already there, along with their respective a priori probabilities, as well as the ability to update them when experience says that they need to be revised. Is this how a baby learns?

CHAPTER 3

Babies' Invisible Knowledge

ON THE SURFACE, WHAT COULD BE MORE DESTITUTE OF KNOWLEDGE than a newborn? What could be more reasonable than to think, as Locke did, that the infant's mind is a "blank slate" simply waiting for the environment to fill its empty pages? Jean-Jacques Rousseau (1712–78) strove to drive this point home in his treatise *Emile, or On Education* (1762): "We are born capable of learning, but knowing nothing, perceiving nothing." Almost two centuries later, Alan Turing, the father of contemporary computer science, took up the hypothesis: "Presumably the child brain is something like a notebook as one buys it from the stationer's. Rather little mechanism, and lots of blank sheets."

We now know that this view is dead wrong—nothing could be further from the truth. Appearances can be deceiving: despite its immaturity, the nascent brain already possesses considerable knowledge inherited from its long evolutionary history. For the most part, however, this knowledge remains invisible, because it does not show in babies' primitive behavior. It therefore took cognitive scientists much ingenuity and significant methodological advances in order to expose the vast repertoire of abilities all babies are born with. Objects, numbers, probabilities, faces, language . . . the scope of babies' prior knowledge is extensive.

THE OBJECT CONCEPT

We all have the intuition that the world is made of rigid objects. In reality, it is made up of atoms, but at the scale on which we live, these atoms are often packed together into coherent entities that move as a single blob and sometimes collide without losing their cohesiveness. . . . These large bundles of atoms are what we call "objects." The existence of objects is a fundamental property of our environment. Is this something that we need to learn? No. Millions of years of evolution seem to have engraved this knowledge into the very core of our brains. As early as a few months of age, a baby already knows that the world is made up of objects that move coherently, occupy space, do not vanish without reason, and cannot be in two different places at the same time.[1] In a sense, babies' brains already know the laws of physics: they expect the trajectory of an object to be continuous in space as in time, without any sudden jump or disappearance.

How do we know this? Because babies act surprised in certain experimental situations that violate the laws of physics. In today's cognitive science laboratories, experimenters have become magicians (see figure 5 in the color insert). In small theaters specially designed for babies, they play all sorts of tricks: on the stage, objects appear, disappear, multiply, pass through walls. . . . Hidden cameras monitor the babies' gazes, and the results are clear-cut: even babies a few weeks old are sensitive to magic. They already possess deep intuitions of the physical world and, like all of us, are stunned when their expectations turn out to be false. By zooming in on the children's eyes—to determine where they look and for how long—cognitive scientists manage to accurately measure their degree of surprise and infer what they expected to see.

Hide an object behind a book, then suddenly crush it flat, as if the hidden object no longer existed (in reality, it escaped through a trapdoor): babies are flabbergasted! They fail to understand that a solid object can vanish into thin air. They appear dumbfounded when an object disappears behind one screen and reappears behind another, without ever being seen in the empty space between the two screens. They are also amazed when a small train rolling down a slope seamlessly passes through a rigid wall. And they expect objects to form a coherent whole: if they see two ends of a stick moving coherently on both sides of a screen, they expect them to belong to a single stick and are shocked when the screen lowers and reveals two distinct rods (see below).

Babies therefore possess a vast knowledge of the world, but they don't know everything from the start, far from it. It takes a few months for babies to understand how two objects can support each other.[2] At first, they don't know that an object falls when you drop it. Only very gradually do they become aware of all the factors that make an object fall or stay put. First, they realize

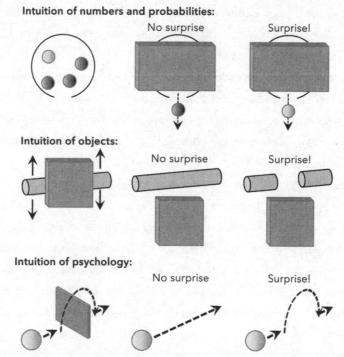

Babies possess extremely early intuitions of arithmetic, physics, and even psychology. To assess them, researchers evaluate whether babies look at a surprising scene for a longer time than an unsurprising one. When a box contains a majority of black balls, babies are surprised to see a white one come out (intuition of numbers and probabilities). If two ends of a stick move coherently, babies are dumbfounded when two different rods are revealed (intuition of objects). And if babies see a ball move autonomously and jump over a wall before escaping to the right-hand side, they deduce that the ball is a living being with an intention of its own, and they are amazed if it keeps jumping once the wall has disappeared (intuition of psychology).

that objects fall when they lose their support, but they think that any sort of contact suffices to keep an object still—for example, when a toy is placed at the edge of a table. Progressively, they realize that the toy must not only be in contact with the table, but on top of it, not under or against it. Finally, it takes them a few more months to figure out that this rule is not enough: in the end, it's the center of gravity of the object that must remain above the table.

Keep this in mind the next time your baby drops his or her spoon from the table for the tenth time, to your great despair: they are only experimenting! Like any scientist, children need a whole series of trials to successively reject all the wrong theories, in the following order: (1) objects stay put in the air; (2) they must touch another object to not fall; (3) they must be atop another object to not fall; (4) the majority of their volume must be above another object to not fall, and so on and so forth.

This experimental attitude continues all the way into adulthood. We are all fascinated with gadgets that seem to violate the usual laws of physics (helium balloons, mobiles in equilibrium, roly-poly toys with a displaced center of gravity . . .), and we all enjoy magic shows where rabbits disappear in a hat and women are sawed in two. These situations entertain us because they violate the intuitions that our brain has held since birth and refined in our first year of life. Josh Tenenbaum, a professor of artificial intelligence and cognitive science at MIT, hypothesizes that babies' brains host a game engine, a mental simulation of the typical behavior of objects similar to the ones that video games use in order to simulate different virtual realities. By running these simulations in their heads, and by comparing simulations with reality, babies discover very early on what is physically possible or probable.

THE NUMBER SENSE

Let's take a second example: arithmetic. What could be more obvious than that babies have no understanding of mathematics? And yet, since the 1980s, experiments have shown quite the opposite.[3] In one experiment, babies are repeatedly presented with slides showing two objects. After a while, they get bored . . . until they are shown a picture with three objects: suddenly, they stare longer at this new scene, indicating that they detected the change. By manipulating the nature, size, and density of objects, one can prove that children are genuinely sensitive to the number itself, i.e., the cardinal of the whole set, not another physical parameter. The best proof that infants possess an abstract "number sense" is that they generalize from sounds to images: if they hear *tu tu tu tu*—that is, four sounds—they are more interested in a picture that has a matching number of four objects in it than in a picture that has twelve, and vice versa.[4] Well-controlled experiments of this sort abound and convincingly show that, at birth, babies already possess the intuitive ability to recognize an approximate number without counting, regardless of whether the information is heard or seen.

Can babies calculate too? Suppose that children see an object hide behind a screen, followed by a second one. The screen then lowers—lo and behold, only one object is there! Babies manifest their surprise in a prolonged investigation of the unexpected scene.[5] If, however, they see the two expected objects, they look at them for only a brief moment. This behavior of "cognitive surprise," in reaction to the violation of a mental calculation, shows that, as early as a few months of age, children understand that 1 + 1 should make 2. They build an internal model of the hidden scene and know how to manipulate it by adding or removing objects. And such experiments work not only for 1 + 1 and 2 − 1, but also for 5 + 5 and 10 − 5. Provided that the error is big enough, nine-month-old babies are surprised whenever a concrete display hints at a wrong calculation: they can tell that 5 + 5 cannot be 5, and that 10 − 5 cannot be 10.[6]

Is this really an innate skill? Could the first months of life suffice for a child to learn the behavior of sets of objects? While children undoubtedly refine the accuracy with which they perceive numbers[7] over the first months of life, the data show, equally clearly, that the starting point for children is not a blank slate. Newborns perceive numbers within a few hours of life—and so

do monkeys, pigeons, ravens, chicks, fish, and even salamanders. And with the
chicks, the experimenters controlled all the sensory inputs, making sure that
the baby chicks did not see even a single object after they hatched . . . yet the
chicks recognized numbers.[8]

Such experiments show that arithmetic is one of the innate skills that evolution bestows unto us, as well as many other species. Brain circuits for numbers
have been identified in monkeys and even in ravens. Their brains contain
"number neurons" that behave in a very similar way: they are attuned to specific numbers of objects. Some neurons prefer to see one object, others two,
three, five, or even thirty objects—and, crucially, these cells are present even
in animals that have not received any specific training.[9] My lab has used brain-
imaging techniques to show that, at homologous locations of the human brain,
our neuronal circuits also contain similar cells attuned to the cardinal number
of a concrete set—and recently, with advances in recording techniques, such
neurons have been directly recorded in the human hippocampus.[10]

Incidentally, these results overturn several tenets of a central theory
of child development, that of the great Swiss psychologist Jean Piaget
(1896–1980). Piaget thought that young infants were not endowed with
"object permanence"—the fact that objects continue to exist when they
are no longer seen—until the end of the first year of life. He also thought
that the abstract concept of number was beyond children's grasp for the first
few years of life, and that they slowly learned it by progressively abstracting
away from the more concrete measures of size, length, and density. In reality,
the opposite is true. Concepts of objects and numbers are fundamental features of our thoughts; they are part of the "core knowledge" with which we
come into the world, and when combined, they enable us to formulate more
complex thoughts.[11]

Number sense is only one example of what I call infants' invisible knowledge: the intuitions that they possess from birth and that guide their subsequent
learning. Here are more examples of the skills researchers have demonstrated
in babies as young as a few weeks old.

THE INTUITION OF PROBABILITIES

Going from numbers to probabilities takes only one step . . . a step that researchers have recently taken by wondering if babies a few months old could predict
the outcome of a lottery draw. In this experiment, babies are first presented
with a transparent box containing balls that move around randomly. There
are four balls: three red and one green. At the bottom, there is an exit. At some
point, the container is occluded, and then either a green ball or a red ball
comes out the bottom. Remarkably, the child's surprise is directly related to
the improbability of what she sees: if a red ball comes out—the most likely
event, since the majority of the balls in the box are red—the baby looks at it
for only a brief moment . . . whereas if the more improbable outcome occurs,
that is, a green ball that had only one chance in four to come out, the baby
looks at it for much longer.

Subsequent controls confirm that babies run, in their little heads, a
detailed mental simulation of the situation and the associated probabilities.
Thus, if we introduce a partition that blocks the balls, or if we move the balls
closer to or farther away from the exit, or if we vary the time before the balls
exit the box, we find that infants integrate all these parameters into their
mental calculation of probability. The duration of their gaze always reflects
the improbability of the observed situation, which they seem to compute
based on the number of objects involved.

All these skills surpass those of most current artificial neural networks.
Indeed, infants' surprise reaction is far from trivial. Being surprised indicates
that the brain was able to estimate the underlying probabilities and concluded
that the observed event had but a tiny chance of occurring. Because babies'
gazes show elaborate signs of surprise, their brains must be capable of probabilistic calculations. Indeed, one of the most popular current theories of brain
function views the brain as a probabilistic computer that manipulates probability distributions and uses them to anticipate future events. Infant experiments reveal that even babies are equipped with such a sophisticated calculator.

A series of recent studies further shows that babies come equipped with all the mechanisms to make complex probabilistic inferences. Do you remember the Reverend Bayes's mathematical theory of probabilities, which allows us to trace an observation back to its probable causes? Well, even babies a few months old already seem to reason according to Bayes's rule.[12] Indeed, not only do they know how to go from a box of colored balls to the corresponding probabilities (forward reasoning) as we just saw, but also from observations back to the content of the box (reverse inference). In one experiment, we first show babies an opaque box, whose contents are hidden. Then we bring in a blindfolded person, who randomly takes out a series of balls. The balls appear one after another, and it turns out that the majority are red. Can babies infer that the box must contain an abundance of red balls? Yes! When we eventually open the box and show them that it contains a majority of green balls, they are surprised and look longer than if the box turns out to be full of red balls. Their logic is impeccable: If the box is filled mostly with green balls, how do we explain that the random draw yielded so many red balls?

Again, this behavior may not seem like much, but it implies an extraordinary ability for implicit, unconscious reasoning that works in both directions: given a sample, infants can guess the characteristics of the set from which it was drawn; and, vice versa, given a set, they manage to guess how a random sample should look.

From birth on, thus, our brain is already endowed with an intuitive logic. There are now many variations of those basic experiments. They all demonstrate the extent to which children behave like budding scientists who reason like good statisticians, eliminating the least likely hypotheses and searching for the hidden causes of various phenomena.[13] For example, the American psychologist Fei Xu showed that if eleven-month-olds see a person draw a majority of red balls from a container, and then find out that the container holds a majority of yellow balls, they are surprised, of course, but they also make an additional inference: that the person prefers the red balls![14] And if they see that a draw is not random but follows a specific pattern, say, a perfect alternation of a yellow ball, a red ball, a yellow ball, a red ball, and so on, then they deduce that a human, not a machine, made the draw.[15]

Logic and probability are closely linked. As Sherlock Holmes put it, "When you have eliminated the impossible, whatever remains, however improbable, must be the truth." In other words, one can turn a probability into a certainty by using reasoning to eliminate some of the possibilities. If a baby can juggle with probabilities, she must also master logic, because logical reasoning is only the restriction of probabilistic reasoning to probabilities 0 and 1.[16] This is exactly what the philosopher and developmental psychologist Luca Bonatti recently showed. In his experiments, a ten-month-old baby first sees two objects, a flower and a dinosaur, hide behind a screen. Then one of these objects exits from the screen, but it is impossible to tell which one because it is partially hidden in a pot, so that only the top can be seen. Later, the dinosaur exits from the other side of the screen, in full sight. At this point, the child can make a logical deduction: "It is either the flower or the dinosaur that is hiding in the pot. But it cannot be the dinosaur, because I have just seen it come out from the other side. So, it must be the flower." And it works: the baby is not surprised if the flower comes out of the pot, but she is if the dinosaur comes out. Furthermore, the baby's gaze reflects the intensity of her logical reasoning: like an adult, her pupils dilate at the precise moment when deduction becomes possible. A true Sherlock Holmes in diapers, the baby seems to start with several hypotheses (it is either the flower or the dinosaur) and then eliminates some of them (it cannot possibly be the dinosaur), thus moving from probability to certainty (it must be the flower).

"Probability theory is the language of science," Jaynes tells us—and infants already speak this language: way before they pronounce their first words, they manipulate probabilities and combine them in sophisticated syllogisms. Their sense of probability allows them to draw logical conclusions from the observations they make. They are constantly experimenting, and their budding scientist brains ceaselessly accumulate the conclusions of their research.

KNOWLEDGE OF ANIMALS AND PEOPLE

While babies have a good model of the behavior of inanimate objects, they also know that there is another category of entities that behave entirely differently: animate things. From the first year of life, babies understand that animals and people have a specific behavior: they are autonomous and driven by their own movements. Therefore, they do not have to wait for another object to bump into them, like a pool ball, in order to move around. Their movement is motivated from within, not caused from the outside.

Babies are therefore not surprised to see animals move by themselves. In fact, for them, any object that moves by itself, even if it is in the shape of a triangle or a square, is immediately labeled as an "animal," and from that moment on, everything changes. A small child knows that living beings do not have to move according to the laws of physics, but that their movements are governed by their intentions and beliefs.

Let us take an example: if we show babies a sphere that moves in a straight line, jumps over a wall, then heads to the right, little by little, they will get bored of it. Are they simply getting used to this peculiar motion? No, in fact, they understand much more. They deduce that this is an animate being with a specific intention: it wants to move to the right! Moreover, they can tell the object is highly motivated, because it jumps over a high wall in order to get there. Now let's remove the wall. In this scenario, babies are not surprised if they see the sphere change its motion and move to the right in a straight line, without jumping—this is simply the best way to attain its goal. On the other hand, babies open their eyes wide if the sphere continues to jump in the air for no particular reason, since the wall has vanished! In the absence of a wall, the same trajectory as in the first scenario leaves the babies surprised, because they do not understand what strange intention the sphere might have.[17] Other experiments show that children routinely infer people's intentions and preferences. In particular, they understand that the higher the wall is, the greater the person's motivation must be in order to jump over it. From their observations, babies can infer not only the goals and intentions of those around them, but also their beliefs, abilities, and preferences.[18]

Infants' notion of living beings does not end there. Around ten months, babies start attributing personalities to people: if they see someone throw a child to the ground, for example, they deduce that this person is ill-intentioned, and they turn away from her. They clearly prefer a second person who helps the child get back up.[19] Long before they are able to pronounce the words *mean* and *nice*, they are able to formulate these concepts in their language of thought. Such a judgment is quite subtle: even a nine-month-old baby can distinguish between someone who intentionally does harm and someone who does it by accident, or someone who intentionally refuses to help another person and someone who does not have the opportunity to help.[20] As we will see later, this social skill plays a fundamental role in learning. Indeed, even a one-year-old child understands if someone is trying to teach him something. He can tell the difference between an ordinary action and an action with the goal of teaching something new. In this respect, a one-year-old child already possesses, according to the Hungarian psychologist György Gergely, an innate sense of pedagogy.

FACE PERCEPTION

One of the earliest manifestations of infants' social skills is the perception of faces. For adults, the slightest hint suffices to trigger the perception of a face: a cartoon, a smiley, a mask. . . . Some people even detect the face of Jesus Christ in the snow or on burnt toast! Remarkably, this hypersensitivity to faces is already present at birth: a baby a few hours old turns its head more quickly to a smiley face than to a similar image turned upside down (even if the experimenter ensures that the newborn has never had the chance to see a face). One team even managed to present a pattern of light to fetuses through the wall of the uterus.[21] Surprisingly, the researchers showed that three dots arranged in the shape of a face (•.•) attracted the fetus more than three dots arranged in the shape of a pyramid (.•.). Face recognition seems to start in utero!

Many researchers believe that this magnetic attraction to faces plays an essential role in the early development of attachment—especially since one of the earliest symptoms of autism is avoiding eye contact. By attracting our eyes to faces, an innate bias would force us to learn to recognize them—and indeed, as early as a couple of months after birth, a region of the visual cortex of the right hemisphere begins to respond to faces more than to other images, such as places.[22] The specialization for faces is one of the best examples of the harmonious collaboration between nature and nurture. In this domain, babies exhibit strictly innate skills (a magnetic attraction to face-like pictures), but also an extraordinary instinct to learn the specifics of face perception. It is precisely the combination of these two factors that allows babies, in a little less than a year, to go beyond naively reacting to the mere presence of two eyes and a mouth and to start preferring human faces to those of other primates, such as monkeys and chimpanzees.[23]

THE LANGUAGE INSTINCT

The social skills of small children are manifest not only in vision, but also in the auditory domain—spoken language comes to them just as easily as face perception. As Steven Pinker famously noted in his best-selling book *The Language Instinct* (1994), "Humans are so innately hardwired for language that they can no more suppress their ability to learn and use language than they can suppress the instinct to pull a hand back from a hot surface." This statement should not be misunderstood: obviously, babies are not born with a full-blown lexicon and grammar, but they possess a remarkable capacity to acquire them in record time. What is hardwired in them is not so much language itself as the ability to acquire it.

Much evidence now confirms this early insight. Right from birth, babies already prefer listening to their native language rather than to a foreign one[24]—a truly extraordinary finding which implies that language learning starts in utero. In fact, by the third trimester of pregnancy, the fetus is already able to hear. The melody of language, filtered through the uterine wall, passes on to babies, and they begin to memorize it. "As soon as the sound of your greeting reached my ears, the baby in my womb leaped for joy," said the pregnant Elizabeth when Mary visited her.[25] The Evangelist was not mistaken: in the last few months of pregnancy, the growing fetus's brain already recognizes certain auditory patterns and melodies, probably unconsciously.[26]

This innate ability is obviously easier to study in premature babies than in fetuses. Out of the womb, we can equip their tiny heads with miniature electroencephalography and cerebral blood flow sensors and peek into their brains. With this method, my wife, professor Ghislaine Dehaene-Lambertz, discovered that even babies born two and a half months before term respond to spoken language: their brain, although immature, already reacts to changes in syllables as well as in voices.[27]

It was long thought that language acquisition does not begin until one or two years of age. Why? Because—as its Latin name, *infans*, suggests—a newborn child does not speak and therefore hides its talents. And yet, in terms of language comprehension, a baby's brain is a true statistical genius. To show this, scientists had to deploy a whole panoply of original methods, including the measurement of infants' preferences for speech and nonspeech stimuli, their responses to change, the recording of their brain signals. . . . These studies gave converging results and revealed how much infants already know about language. Right at birth, babies can tell the difference between most vowels and consonants in every language in the world. They already perceive them as categories. Take, for instance, the syllables /ba/, /da/, and /ga/: even if the corresponding sounds vary continuously, babies' brains treat them as distinct categories separated by sharp borders, just like adults.

These early innate skills become shaped by the linguistic environment during the first year of life. Babies quickly notice that certain sounds are not used in their language: English speakers never utter vowels like the French /u/ and /eu/, and Japanese locutors fail to differentiate between /R/ and /L/. In just a few months (six for vowels, twelve for consonants), the baby's brain sorts through its initial hypotheses and keeps only the phonemes that are relevant to the languages that are present in its environment.

But that's not all: babies quickly start to learn their first words. How do they go about identifying them? First, babies rely on prosody, the rhythm and intonation of speech—the way our voices rise, fall, or stop, thus marking the boundaries between words and sentences. Another mechanism identifies which speech sounds follow each other. Again, babies behave like budding statisticians. They realize, for example, that the syllable /bo/ is often followed by /t^l/. A quick calculation of probabilities tells them that this cannot be due to chance: /t^l/ follows /bo/ with too high a probability; these syllables must form a word, "bottle"—and this is how this word is added to the child's vocabulary and can later be related to a specific object or concept.[28] As early as six months of age, children have already extracted the words that recur with a high frequency in their environment, such as "baby," "daddy," "mommy," "bottle," "foot," "drink," "diaper," and so forth. These words become engraved in their memory to such an extent that, as adults, they continue to hold a special status and are processed more effectively than other words of comparable meaning, sound, and frequency acquired later in life.

Statistical analysis also allows babies to identify certain words that occur more frequently than others: small grammatical words such as articles (a, an, the) and pronouns (I, you, he, she, it . . .). By the end of their first year, babies already know many of them, and they use them to find other words. If, for example, they hear one of their parents say, "I made a cake," they can parse out the small function words "I" and "a" and, by elimination, discover that "made" and "cake" are also words. They already understand that a noun often comes after an article and a verb usually comes after a pronoun—to such an extent that, around twenty months of age, babies react with utter surprise if they are told incoherent phrases like "I bottle" or "the finishes."[29]

Of course, such a probabilistic analysis isn't entirely foolproof. When French children hear "*un avion*" (an airplane), which is pronounced with a liaison (the n of "un" melds into the a of "avion"), they improperly infer the existence of the word *navion* (*"Regarde le navion!"*). Conversely, English speakers imported the French word *napperon* (place mat) and, due to incorrect parsing of the phrase *un napperon*, invented the word *apron*.

Such shortcomings are rare, however. In a few months, children quickly manage to surpass any existing artificial intelligence algorithm. By the time they blow out their first candle, they have already laid down the foundation for the main rules of their native language at several levels, from elementary sounds (phonemes) to melody (prosody), vocabulary (lexicon), and grammar rules (syntax).

No other primate species is capable of such abilities. This very experiment has been attempted many times: several scientists tried adopting baby chimpanzees, treating them like family members, speaking to them in English or sign language or with visual symbols . . . only to find out, a few years later, that none of these animals mastered a language worthy of the name: they knew, at most, a few hundred words.[30] The linguist Noam Chomsky, therefore, was probably right in postulating that our species is born with a "language acquisition device," a specialized system that is automatically triggered in the first years of life. As Darwin said in *The Descent of Man* (1871), language "certainly is not a true instinct, for every language has to be learnt," but it is "an instinctive tendency to acquire an art." What is innate in us is the instinct to learn any language—an instinct so irrepressible that language appears spontaneously within a few generations in humans deprived of it. Even in deaf communities, a highly structured sign language, with universal linguistic characteristics, emerges from the second generation on.[31]

CHAPTER 4

The Birth of a Brain

The child is born with an unfinished brain and not, as the postulate of the old pedagogy affirmed, with an unoccupied brain.

Gaston Bachelard, *The Philosophy of No:*
A Philosophy of the New Scientific Mind (1940)

Genius without education is like silver in the mine.

Benjamin Franklin (1706-1790)

THE FACT THAT NEWBORN BABIES IMMEDIATELY EXHIBIT SOPHISTICATED knowledge of objects, numbers, people, and languages refutes the hypothesis that their brains are nothing but blank slates, sponges that absorb whatever the environment imposes on them. A simple prediction ensues: if we could dissect the brain of a newborn, we should observe, at birth and perhaps even earlier, well-organized neuronal structures corresponding to each of those major domains of knowledge.

This idea has long been contested. Until about twenty years ago, the newborn's brain was terra incognita. Brain imaging had just been invented—it had not been applied to developing brains yet—and the predominant theoretical

vision was that of empiricism, the idea that the brain is born void of all knowledge, influenced only by its environment. It was only with the advent of sophisticated magnetic resonance imaging (MRI) methods that we were finally able
to visualize the early organization of the human brain and discover that, in
agreement with our expectations, virtually all the circuits of the adult brain
are already present in that of a newborn baby.

THE INFANT BRAIN IS WELL ORGANIZED

My wife, Ghislaine Dehaene-Lambertz, and I, along with our neurologist
colleague Lucie Hertz-Pannier, were among the first to use functional MRI
in two-month-old babies.[1] Of course, we heavily relied on the previous experience of pediatricians. Fifteen years of clinical experience had convinced
them that MRI was a harmless exam that one could prescribe to individuals
of any age, including premature infants. However, practitioners resorted to
this technology only for diagnostic purposes, in order to detect early lesions.
No one had used functional MRI in normally developing babies to see if their
brain circuits could be selectively activated to certain stimuli. To achieve
this, we had to overcome a whole series of difficulties. We designed a noise-
reducing helmet to protect the babies from the loud noise of the machine; we
kept them still by snuggly swaddling them in a cradle made to fit the shape of
the MRI coil; we reassured them by progressively acclimating them to the
unusual environment; and we permanently kept an eye on them from outside
the machine.

In the end, our efforts were rewarded with spectacular results. We had
chosen to focus on language because we knew that babies began to learn it
very quickly over the course of their first year of life. And indeed, we observed
that at two months after birth, when babies heard sentences in their native
language, they activated the very same regions of the brain as adults (see
figure 6 in the color insert).

When we hear a sentence, the first region of the cortex to activate is the primary auditory area—this is the entry point for all auditory information into the brain. This area also lit up in the infant brain as soon as the sentence began. This may seem obvious to you, but at the time, it was not self-evident for very young infants. Some researchers presumed that the sensory areas of children's brains are so disorganized at birth that their senses tend to blend. According to these researchers, for several weeks, a baby's brain mixes hearing, vision, and touch, and it takes some time for the baby to learn to separate these sensory modalities.[2] We know today that this is false—from birth on, hearing activates the auditory areas, vision activates the visual areas, and touch activates the areas associated with tactile sensation, without us ever having to learn this. The subdivision of the cortex into distinct territories for each of the senses is given to us by our genes. All mammals possess it, and its origin is lost in the arborescence of our evolution (see figure 7 in the color insert).[3]

But let's go back to our experiment where babies listened to sentences in the MRI scanner. After entering the primary auditory area, the activity spread rapidly. A fraction of a second later, other areas lit up, in a fixed order: first, the secondary auditory regions, adjacent to the primary sensory cortex; then a whole set of temporal lobe regions, forming a gradual stream; and finally Broca's area, at the base of the left frontal lobe, simultaneously with the tip of the temporal lobe. This sophisticated information processing chain, lateralized to the left hemisphere, is remarkably similar to that of an adult. At two months, babies already activate the same hierarchy of phonological, lexical, syntactic, and semantic brain areas as adults. And, just as in adults, the more the signal climbs up in the hierarchy of the cortex, the slower the brain responses are and the more these areas integrate information on an increasingly high level (see figure 6 in the color insert).[4]

Of course, two-month-old babies do not yet understand the sentences that they hear; they have yet to discover words and grammatical rules. However, in their brains, linguistic information is channeled into highly specialized circuits similar to those of adults. Babies learn to understand and speak so quickly—while all other primates are unable to do so—probably because their left hemisphere comes equipped with a predetermined hierarchy of circuits that specialize in detecting statistical regularities for all aspects of speech: sound, word, sentence, and text.

LANGUAGE HIGHWAYS

Activity flows through all these brain areas in a specific order because they are connected to one another. In adults, we are beginning to understand which neuronal pathways interconnect the language regions. In particular, neurologists have discovered that a large cable made of millions of nerve fibers, called the "arcuate fasciculus," connects the temporal and parietal language areas at the back of the brain with frontal areas, notably the famous Broca's area. This bundle of connections is a marker of the evolution of language. It is much larger in the left hemisphere, which, in 96 percent of right-handers, is devoted to language. Its asymmetry is specific to humans and is not observed in other primates, not even our closest cousins, the chimpanzees.

Once again, this anatomical characteristic is not the outcome of learning: it is there from the start. In fact, when we examine the connections of a newborn's brain, we discover that not only the arcuate fasciculus but all the major fiber bundles that connect cortical and subcortical brain areas are in place at birth (see figure 8 in the color insert).[5]

These "highways of the brain" are built during the third trimester of pregnancy. During the construction of the cortex, each growing excitatory neuron sends out its axon to explore the surrounding regions, sometimes up to several centimeters away, like a Christopher Columbus of the brain. This exploration is guided and channeled by chemical messages, molecules whose concentrations vary from one region to another and which act as spatial labels. The axon head literally sniffs out this chemical path laid out by our genes, and deduces the direction in which it must go. Thus, without any intervention from the outside world, the brain self-organizes into a network of crisscrossed connections, several of which are specific to the human species. As we will see in a moment, this network can be further refined by learning—but the initial scaffolding is innate and built in utero.

Should we be surprised? Only twenty years ago, many researchers considered it extremely unlikely that the brain was anything else but a disorganized mass of random connections.[6] They could not imagine that our DNA, which contains only a limited number of genes, could host a detailed blueprint of the highly specialized circuits that support vision, language, and motor skills. But this is improper reasoning. Our genome contains all the details of our body: it knows how to make a heart with four chambers; it routinely constructs two eyes, twenty-four vertebrae, the inner ear and its three perpendicular channels, ten fingers and their phalanges, all with extreme reproducibility . . . so why not a brain with multiple internal subregions?

Recent advances in biological imaging have revealed how, as early as the first two months of pregnancy, when the fingers of the hand are still only buds, they are already invaded by three nerves, the radial, the median, and the ulnar, each targeting specific end points (see figure 8 in the color insert).[7] The same high-precision mechanics may therefore exist in the brain: just as the bud of the hand splits into five fingers, the cortex subdivides into several dozen highly specialized regions separated by sharp borders (see figure 9 in the color insert).[8] As early as the first months of pregnancy, many genes are selectively expressed at different points in the cortex.[9] Around twenty-eight weeks of gestation, the brain begins to fold, and the main sulci that characterize the human brain appear. In thirty-five-week-old fetuses, all the major folds of the cortex are well formed, and the characteristic asymmetry of the temporal region, which houses the language areas, can already be seen.[10]

THE SELF-ORGANIZATION OF THE CORTEX

Throughout pregnancy, as cortical connections develop, so do the corresponding cortical folds. In the second trimester, the cortex is initially smooth; then, a first set of ridges emerges, reminiscent of the monkey brain; and finally, we begin to see the secondary and tertiary folds typical of the human brain—folds upon folds upon folds. Their epigenesis gradually becomes more and more dependent on the activity of the nervous system. Depending on the feedback that the brain receives from the senses, some circuits stabilize, while others, useless, degenerate. Thus, the folding of the motor cortex ends up being slightly different in left-handers and right-handers. Interestingly, left-handed individuals who were forced to write with their right hand as children show a sort of compromise: the shape of their motor cortex is typical of a left-hander, but its size exhibits the left-right asymmetry of a right-handed person.[11] As the authors of this study conclude, "cortical morphology in adults holds an accumulated record of both innate biases and early developmental experience."

Cortical folds in the fetus's brain owe their spontaneous formation to a biochemical self-organization process that depends on both the genes and the chemical environment of the cells, requiring extremely little genetic information and no learning at all.[12] Such self-organization isn't nearly as paradoxical as it sounds—in fact, it is omnipresent on earth. Picture the cortex as a sandy beach on which ripples and pools form, at multiple scales, as the tides come and go. Or imagine it as a desert in which wrinkles and dunes appear under the relentless action of the wind. In fact, stripes, spots, and hexagonal cells emerge in all kinds of biological or physical systems on many scales, from fingerprints to zebra skin, leopard spots, basalt columns in volcanoes, desert dunes, and regularly spaced clouds in a summer sky. The British mathematician Alan Turing was the first to explain this phenomenon: all that is needed is a process of local amplification and inhibition at a distance. When the wind blows over a beach, as grains of sand begin to accumulate, a self-amplification process begins: the emerging bump tends to catch other grains of sand, while behind it, the wind swirls and tears away at the sand; after a few hours, a dune is born. As soon as there is local excitation and inhibition at a distance, we can see a dense region (the dune) appear, surrounded by a less dense region (the hollow side), itself followed by another dune, ad infinitum. Depending on the exact circumstances, the patterns that spontaneously emerge form spots, stripes, or hexagons.

Self-organization is ubiquitous in the developing brain: our cortex is full of columns, stripes, and sharp borders. Spatial segregation seems to be one of the mechanisms by which the genes lay out neuronal modules specialized for processing different types of information. The visual cortex, for example, is covered with alternating bands that process information from the left and right eyes—they are called "ocular dominance columns" and they emerge spontaneously in the developing brain, using the information arising from intrinsic activity in the retina. But similar mechanisms of self-organization can occur at a higher level, not necessarily to tile the surface of the cortex, but to cover more abstract space. One of the most spectacular examples is the existence of *grid cells*—neurons that encode the location of a rat by paving space with a grid of triangles and hexagons (see figure 10 in the color insert).

Grid cells are neurons located in a specific region of the rat brain called the "entorhinal cortex." Edvard and May-Britt Moser earned the Nobel Prize in 2014 for discovering their remarkable geometrical properties. They were the first to record from neurons in the entorhinal cortex while the animal moved around in a very large room.[13] We already knew that in a nearby region called the "hippocampus," neurons behaved as "place cells": they fired only if the animal was in a specific location in the room. The Moser's groundbreaking discovery was that grid cells did not respond to just a single place, but to a whole set of positions. Furthermore, those privileged locations which made a given cell fire were regularly arrayed: they formed a network of equilateral triangles that grouped together to form hexagons, a bit like the spots on the skin of a giraffe or the basalt columns in volcanic rocks! Whenever the animal walks around, even in darkness, the firing of each grid cell tells the rat where it is in relation to a network of triangles that span the entire space. The Nobel committee rightly called this system the "GPS of the brain": it provides a highly reliable neuronal coordinate system that maps external space.

But why do neuronal maps use triangles and hexagons, rather than the rectangles and perpendicular lines of our usual charts? Since Descartes, mathematicians and cartographers have always relied on two perpendicular axes called "Cartesian coordinates" (x and y, abscissa and ordinate, longitude and latitude). Why does the rat brain prefer to rely on a set of triangles and hexagons? Most likely because the grid-cell neurons self-organize during development— and in nature, such self-organization frequently produces hexagons, from giraffe skin to beehives to volcanic columns. Physicists now understand why hexagonal shapes are so ubiquitous: they spontaneously emerge whenever a system starts from a disorganized "hot" state and slowly cools down, eventually freezing into a stable structure (see figure 10 in the color insert). Researchers have proposed a similar theory for the emergence of grid cells in the entorhinal cortex during brain development: disorganized groups of neurons would progressively settle down into an organized set of grid cells, with hexagons emerging as a sponta- neous attractor of the dynamics of the cortex.[14] According to this theory, no teaching signal is required for the rat to grow a grid-like map. In fact, the setting up of this circuit does not involve any learning at all: it naturally emerges from the dynamics of the developing cortex.

This theory of the self-organization of brain maps is beginning to be successfully tested. Extraordinary experiments show that the brain's GPS does indeed emerge very early on during rat development. Two independent groups of researchers succeeded in implanting electrodes in baby rats, barely born, before they even started to walk.[15] Using this setup, they examined whether grid cells were already present in the entorhinal cortex. They also probed place cells (those that respond to a single location) and head direction cells, a third type of neuron that functions like a ship's compass: each neuron fires when the animal moves in a certain direction, for example, northwest or southeast. What the researchers found is that this whole system is practically innate: the head direction cells are present as soon as one can record, and the place and grid cells emerge one or two days after the baby rat has begun mov- ing around.

The data are lovely, but they should hardly be surprising: for most ani- mals, from ants to birds, reptiles, and mammals, map making is a major mat- ter. As soon as pups, kittens, or babies leave the nest and explore the world, it is crucial to their survival that they know where they are at all times and can find their way home, where their mothers await them. Eons ago, evolution seems to have hit upon a way to provide the nascent brain with a compass, a map, and a record of the places it visits.

Indeed, does this neuronal GPS exist in the human brain? Yes. We now know, by indirect means, that the adult brain also contains a neuronal map with hexagonal symmetry, in exactly the same place as in rats (the entorhinal cortex).[16] And we also know that very small children already have a sense of space. Toddlers have no difficulty orienting themselves in a room: if they are taken from point A to point B, then to point C, they will know how to return in a straight line from C to A—and remarkably, they do so even if they are blind from birth. The young of the human species thus possess, like rats, a mental module for spatial navigation.[17] We have not yet managed to directly see this map in the baby's brain, because it remains extraordinarily difficult to obtain images of a brain in action at this very young age (try doing an MRI on a crawling baby). But we are pretty sure that we will find it one day, as soon as mobile brain-imaging methods become available.

I could go on and on about the examples of other specialized modules in a baby's brain. We know, for example, that as early as a few months of age (although perhaps not at birth), the visual cortex contains a region that responds preferentially to faces, more so than to images of houses.[18] The formation of this region seems to be partly the result of learning, but it is tightly channeled, guided, and constrained by the brain's connectivity. Those connections ensure that the same location, give or take a few millimeters, specializes for faces in all individuals—it ends up forming one of the most specific modules of the cortex, a patch where up to 98 percent of neurons specialize for faces and barely respond to other pictures.

To take another example, we also know that a baby's parietal cortex already responds to the number of objects,[19] at a location that matches the region which is activated when a human adult calculates 2 + 2, or when a monkey memorizes a number of objects. In monkeys, the German neuroscientist Andreas Nieder successfully demonstrated that this region contains neurons sensitive to the number of objects: there are specialized neurons for one object, others for two objects, three objects, and so on and so forth . . . and these neurons are present even if the monkey in question has never been trained to perform a numerical task. We therefore think that these modules initially emerge on an innate basis, even if the environment later shapes them. My colleagues and I have proposed a precise mathematical model for the self-organization of number neurons, this time based on a wavelike propagation of activity along the surface of the developing cortex. This theory can explain the properties of number neurons in every detail. In the model, these cells end up forming a sort of number line—a linear chain that spontaneously emerges out of a network of randomly connected neurons in which the numbers one, two, three, four, and so on occupy successive positions.[20]

The concept of self-organization departs radically from the classical—but wrong—view of the brain as a blank slate, largely devoid of initial structure and dependent on the environment to configure it. Contrary to this view, little or no data is needed for the brain to grow a map or a number line. Self-organization also sets the brain apart from the artificial neural networks that currently dominate the engineering approaches to artificial intelligence. Nowadays, AI has become virtually synonymous with big data—because those networks are incredibly data-hungry and begin to act intelligently only after they have been fed with gigabytes of data. Unlike them, however, our brain does not require so much experience. Quite the contrary, the main nodes of our brain, the modules where our core knowledge is stored, seem to develop largely spontaneously, perhaps purely through internal simulation.

Only a handful of contemporary computer scientists, such as MIT professor Josh Tenenbaum, are seriously attempting to incorporate this type of self-organization into artificial intelligence. Tenenbaum and his colleagues are working on the "virtual baby project"—a system that would come into the world with the ability to self-generate millions of thoughts and images. These internally generated data would then serve as a basis for learning in the rest of the system, without the need to provide any additional external data. According to this radical vision, even before birth, the foundations of our core brain circuits arise through self-organization, by bootstrapping themselves from a database generated inside the system.[21] Most of the initial groundwork occurs internally, in the absence of any interaction with the outside world; only the final adjustments are left to learning, shaped by the additional data that we receive from our environment.

The conclusion that emerges from this line of research emphasizes the joint power of genes and self-organization in the development of the human brain. At birth, the baby's cortex is folded almost like an adult's. It is already subdivided into specialized sensory and cognitive areas, which are interconnected by precise and reproducible bundles of fibers. It hosts a collection of partially specialized modules, each of which projects a particular type of representation onto the outside world. The grid cells of the entorhinal cortex draw two-dimensional planes, perfect for coding and navigating space. As we will see later, other regions, such as the parietal cortex, draw lines, excellent for coding linear quantities including number, size, and the passing of time; and Broca's area projects tree structures, ideal for coding the syntax of languages. From our evolution, we inherit a set of fundamental rules from which we will later select those that best represent the situations and concepts that we will have to learn in our lifetime.

THE ORIGINS OF INDIVIDUALITY

By asserting the existence of a universally human nature, an innate brain circuitry laid out by genes and self-organization, I do not mean to deny the existence of individual differences. Wherever we zoom in, each of our brains exhibits unique traits—even right from the start. For instance, our cortical folds, much like our fingerprints, are laid down prior to birth and vary in distinctive ways—even in identical twins. Similarly, the strength and density of our long-distance cortical connections, and even their exact trajectories, vary by a large factor and make each of our "connectomes" unique.

It is important to recognize, however, that these variations ride on a common theme. The layout of the *Homo sapiens* brain obeys a fixed scheme, similar to the succession of chords that jazz musicians memorize when they learn a song. It is only on top of this human-universal grid that the vagaries of our genomes and the quirks of our pregnancies add their personal improvisations. Our individuality is real, but it should not be exaggerated: each of us is but a variation on the *Homo sapiens* melody line. In any of us, black or white, Asian or Native American, anywhere on the planet, the human brain architecture is always obvious. In that respect, the cortex of any human differs from that of our closest living relative, the chimpanzee, just as much as any improvisation on "My Funny Valentine" departs from, say, one on "My Romance."

Because we all share the same initial brain structure, the same core knowledge, and the same learning algorithms that allow us to acquire additional talents, we often end up sharing the same concepts. The same human potential is present in every person—be it for reading, science, or mathematics, and whether we are blind, deaf, or mute. As the British philosopher Roger Bacon (1220–92) observed in the thirteenth century, "The knowledge of mathematical things is almost innate in us. . . . This is the easiest of sciences, a fact which is obvious in that no one's brain rejects it; for laymen and people who are utterly illiterate know how to count and reckon." The same, obviously, could be said of language—there is virtually no child who does not have the powerful innate drive to acquire the language of its surroundings, whereas, as noted earlier, no chimpanzee, even those adopted by human families at birth, ever mutters more than a few words or composes more than a few signs.

In brief, individual differences are real—but they are almost always of degree rather than of kind. It is only at the extremes of the normal distribution of brain organization that neurobiological variations end up making a real cognitive difference. Increasingly, we are discovering that children with developmental disorders lie on the ends of this distribution. Their brains seem to have taken a wrong turn on the developmental path that leads from genetic inheritance to neuronal migration and circuit self-organization during pregnancy.

The scientific demonstration is increasingly solid in the case of dyslexia, a specific developmental disorder that affects the ability to learn to read while leaving intelligence and other faculties intact. If you are dyslexic, then any of your siblings has a 50 percent chance of also suffering from dyslexia, thus pointing to the strong genetic determinism of this developmental disorder. At least four genes have now been implicated in dyslexia—and interestingly, most of these genes affect the ability of neurons to migrate to their final locations in the cortex during pregnancy.[22] Magnetic resonance also shows profound anomalies in the connections that support reading in the left hemisphere.[23] Crucially, anomalies can be found early on: in children with a genetic predisposition for dyslexia, at six months of age, a deficit in distinguishing the phonemes of spoken language already separates those who will develop dyslexia from those who will turn into normal readers.[24] Indeed, phonological deficits are known to be a major factor in the emergence of dyslexia—but they are not the only cause: the reading circuit is complicated enough that there are many places where it may fail. Various types of dyslexias have been described, including attention deficits that cause the child to mix up the letters in nearby words[25] and visual deficits that cause mirror confusions.[26] Dyslexia seems to lie on the extreme of a bell-curve continuum of visual, attentional, and phonological abilities, ranging from full normality to severe deficit.[27] We all share the same *Homo sapiens* makeup, but we differ slightly in the quantitative amount of our inheritance, probably due to semi-random variations in the early layout of our neural circuits.

Virtually the same story could be told of other developmental deficits. Dyscalculia, for instance, has been related to early gray- and white-matter deficits in the dorsal parietal and frontal circuits that support calculation and mathematics.[28] Premature children, who may suffer from periventricular infarcts in the parietal region supporting number sense, are at a greater risk of dyscalculia.[29] Early neurological disorganization may cause dyscalculia either by directly impacting the core knowledge of sets and quantities, or by disconnecting it from other areas involved in acquiring number words and the symbols of arithmetic. In either case, the outcome is a predisposition for childhood difficulties in acquiring math. Those children are likely to require specific help to strengthen their weak initial intuitions of quantities.

With our black-and-white minds, we tend to exaggerate the consequences of these scientific discoveries on the genetic underpinnings of developmental deficits. None of the genes that are involved in dyslexia, dyscalculia, or, for that matter, any other developmental syndrome, including autism and schizophrenia, have 100 percent determinism. At most, they strongly tip the scales—but the environment also has a huge share in the developmental trajectory that a child will ultimately take. My colleagues in special education are positive: with enough effort, no dyslexia or dyscalculia is so strong as to be beyond the reach of rehabilitation. It is high time that we now turn to this second major player in brain development: brain plasticity.

CHAPTER 5

Nurture's Share

Everybody knows that the ability of a pianist ... requires many years of mental and muscular gymnastics. To understand this important phenomenon, it is necessary to accept that, in addition to the reinforcement of pre-established organic pathways, new pathways are created by the ramification and progressive growth of terminal dendritic and axonal processes.

Santiago Ramón y Cajal (1904)

I HAVE JUST INSISTED ON THE CONTRIBUTION OF NATURE TO THE CON-struction of our brain—the interplay of genes and self-organization. But, of course, nurture is equally important. The brain's early organization does not remain forever unchanged: experience refines and enriches it. This is the other side of the coin: How does learning change the circuits in a child's brain? To understand this, we have to rewind the clock a century, back to the fundamental discoveries of the great Spanish anatomist Santiago Ramón y Cajal (1852–1934).

Cajal is one of the heroes of neuroscience. With his microscope in hand, he was the first to map the micro-organization of the brain. A genius draftsman, he produced realistic yet simplified drawings of neural circuits, true master-pieces that figure among the major works of scientific illustration. But above all, he was able to move from observation to interpretation and from anatomy to function with impressive judgment. Although his microscope showed him only the postmortem anatomy of neurons and their circuits, he nevertheless managed to draw bold and accurate inferences about the way they function.

Cajal's greatest discovery, for which he earned a Nobel Prize in 1906, was that the brain is made up of distinct nerve cells (neurons) and not a continu-ous network, a reticulum, as was previously thought. He also realized that, unlike most other cells—such as red blood cells, which are roughly round and compact—neurons assume incredibly complex shapes. Each neuron has a huge tree composed of several thousand branches, each one smaller than the next, called "dendrites" (*dendron* means "tree" in Greek). Populations of neurons assemble to form an inextricable forest of neuronal arborizations.

This complexity did not discourage our Spanish neuroscientist. In dia-grams that have remained famous in the history of neuroscience, and which depicted the detailed anatomy of the cortex and hippocampus, Cajal added something eminently simple yet luminous and of great theoretical signifi-cance: arrows! Cajal's arrows indicate the direction in which nerve impulses flow: from the dendrites to the cell body of the neuron and, finally, along the axon. It was a bold speculation, but it turned out to be right. Cajal understood that the shape of neurons corresponds to their function: with its dendritic tree, a neuron collects information from other cells, and all those messages con-verge in the cell's body, where the neuron compiles them to send out only a single message. That message, called the "action potential" or "spike," is then transferred along the axon, a long ivy-like liana that reaches out to thousands of other neurons, sometimes several centimeters away.

Cajal was able to infer another point of utmost importance: that neurons communicate with one another through synapses. He was the first to under-stand that each neuron is a distinct cell—but his microscope also revealed that these cells come into contact at certain points. These junction zones are

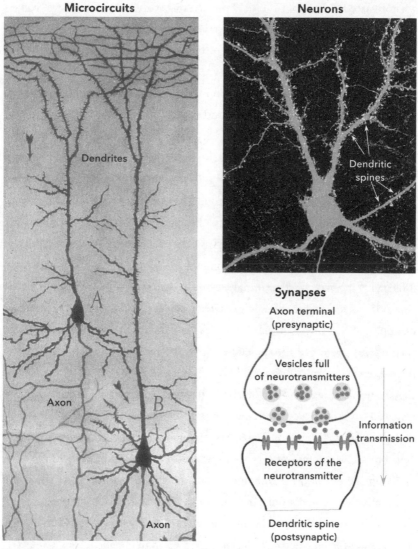

Microcircuits

Dendrites

A

Axon

B

Axon

Neurons

Dendritic spines

Synapses

Axon terminal
(presynaptic)

Vesicles full
of neurotransmitters

Information
transmission

Receptors of the
neurotransmitter

Dendritic spine
(postsynaptic)

Neurons, synapses, and the microcircuits that they form are the material hardware of brain plasticity: they adjust each time we learn. Each neuron is a distinct cell with "trees," called "dendrites" (top left), that collect information from other neurons, and an axon (bottom left) that sends messages to other neurons. A microscope easily resolves the dendritic spines, which are the mushroom-shaped bodies that harbor the synapses—the points of connection between two neurons. As we learn, all these elements can change: the presence, number, and strength of synapses; the size of dendritic spines; the number of branches of dendrites and axons; and even the number of sheets of myelin, which insulates axons and determines their transmission speed.

what we now call "synapses" (Cajal made the discovery, but the name was coined in 1897 by the great British physiologist Charles Sherrington [1857–1952]). Each synapse is the meeting point of two neurons or, more precisely, the place where the axon of one neuron meets the dendrite of another neuron. A "presynaptic" neuron sends its axon far away until it meets the dendrite of a second, "postsynaptic" neuron and connects to it.

What happens at a synapse? Another Nobel Prize winner, the neurophysiologist Thomas Südhof, devoted all his research to this question, and he concluded that synapses are the computing units of the nervous system—genuine nanoprocessors of the brain. Keep in mind that our brain contains about a thousand trillion synapses. The complexity of such machinery is truly unequalled. Here, I can summarize only its simplest features. The message that travels in the axon is electrical, but most synapses transform it into a chemical one. The end of the axon, the "terminal button" near the synapse, contains vesicles, tiny pockets filled with molecules called "neurotransmitters" (glutamate, for example). When the electrical signal reaches the terminal button of an axon, the vesicles open, and the molecules flow into the synaptic space that separates the two neurons. This is why we call these molecules neurotransmitters: they transmit a message from one neuron to the next. A moment after they are released from the presynaptic terminal, the molecules attach themselves to the membrane of the second, postsynaptic neuron, at particular points called "receptors." Neurotransmitters are to receptors like keys are to locks: they literally open doors in the membrane of the postsynaptic neuron. Ions, positively or negatively charged atoms, pour into those open channels and generate an electric current within the postsynaptic neuron. The cycle is complete: the message went from electrical to chemical, then from chemical back to electrical, and in the process, it crossed the space between the two neurons.

What does this have to do with learning? Well, our synapses are constantly changing, throughout our lives, and these changes reflect what we learn.[1] Each synapse is a small chemical plant, and many elements of this plant can change in the course of learning: the number of vesicles, their size, the number of receptors, their efficiency, and even the size and shape of the synapse itself. . . . All these parameters affect the strength with which the presynaptic electrical message will be transmitted to the second, postsynaptic neuron—and they provide useful storage space for learned information.

Moreover, these changes in synaptic strength do not happen at random: they tend to stabilize the activity of neurons, by reinforcing their ability to excite one another if they have already done so in the past. The basic rule is so simple that it was already hypothesized in 1949 by psychologist Donald Hebb (1904–85). It can be summed up in a simple formula: *neurons that fire together, wire together.* When two neurons are activated at the same time or in short succession, their connection strengthens. More precisely, if the emitting presynaptic neuron fires and the postsynaptic neuron fires a few milliseconds later, then the synapse is strengthened: in the future, transmission between these two neurons will be even more efficient. If, on the other hand, the synapse fails to make itself heard, so that the postsynaptic neuron fails to fire, then the synapse weakens.

We now understand why this phenomenon stabilizes neuronal activity: it strengthens circuits that have worked well in the past. Synaptic changes that follow Hebb's rule enhance the probability that the same type of activity happens again. Synaptic plasticity enables vast neuronal tapestries, each composed of millions of neurons, to follow one another in a precise and reproducible order. A mouse that traverses a maze along the optimal path, a violinist who pours a fountain of notes out of her fingers, or a child who successfully recites a poem . . . each of these scenarios awakens a neural symphony in which every move, note, or word is recorded by hundreds of millions of synapses.

Of course, the brain does not keep a record of every event of our lives. Only the moments that it considers the most important get imprinted in our synapses. To this aim, synaptic plasticity is modulated by vast networks of neurotransmitters, particularly acetylcholine, dopamine, and serotonin, that signal which episodes are important enough to remember. Dopamine, for example, is the neurotransmitter associated with reward: food, sex, drugs . . . and in case you were wondering: yes, even rock 'n' roll![2] The dopamine circuit flags everything we love, every stimulus we are "addicted to," and signals to the rest of the brain that what we experience is positive and better than we expected. Acetylcholine, on the other hand, attaches itself more generally to all important moments. Its effects are massive. For example, you are able to remember exquisite details of what you were doing on September 11, 2001, when you learned about the World Trade Center attack, because on that day, a hurricane of neurotransmitters rushed through your brain circuits, causing your synapses to be massively altered. One circuit is particularly crucial: the amygdala, a subcortical group of neurons triggered primarily by strong emotions, sends signals to the nearby hippocampus, which stores the major episodes of our existence. In this manner, synaptic modifications primarily highlight the facts of our lives that the emotional circuits of our brain consider the most significant.

The ability of synapses to modify themselves according to the activity of their pre- and postsynaptic neurons was initially discovered under artificial conditions. Experimenters had to tetanize the neurons by stimulating them at a frantic rate with a strong electric current before the strength of their synapses changed. After this traumatic experience, the synapses remained modified for several hours, a phenomenon called "long-term potentiation," which seemed ideal to maintain memories in the long term.[3] But was this mechanism genuinely used by the brain to store information under normal conditions? The first evidence came from a marine animal, *Aplysia californica*, a sea slug with gigantic neurons. This creature is not endowed with a brain in the typical sense of the word, but rather has large bundles of nerve cells, called "ganglia." In these structures, Nobel Prize winner Eric Kandel identified a whole cascade of synaptic and molecular modifications when the animal became conditioned to expect food, a bit like Pavlov's dog.[4]

Soon, as synapse recording and visualization techniques evolved, evidence accumulated in support of the role of synaptic plasticity in learning. Synaptic changes occur precisely in the circuits that the animal uses to learn. When a mouse learns to avoid a place where it received a small electrical shock, the synapses of the hippocampus, the region responsible for spatial and episodic memory, change:[5] the connections between the hippocampus and the amygdala hardwire such a traumatic experience. When the mouse becomes terrified by a sound, the synapses that connect the amygdala to the auditory cortex undergo a similar change.[6] Furthermore, these changes do not simply co-occur during learning: they actually seem to play a causal role in it. The proof is that if, in the minutes following a traumatic event, we interfere with the molecular mechanisms that allow the synapses to undergo learning-related changes, the animal ends up not remembering anything.[7]

THE PORTRAIT OF A MEMORY

What is a memory? And what is its physical foundation in the brain? Most researchers agree on the following explanation, which distinguishes between periods of encoding and remembering.[8]

Let's start with encoding. Each of our perceptions, actions, and thoughts relies on the activity of a specific subset of neurons (while others remain inactive or even inhibited). The identity of these active neurons, distributed in many regions of the brain, defines the content of our thoughts. When I see, let's say, Donald Trump in the Oval Office, some neurons respond to his face (in the inferior temporal region), others to his voice (in the superior temporal region), others to the layout of his office (in the parahippocampal region), and so on and so forth. Single neurons may provide some information, but the overall memory is always encoded by several interconnected groups of neurons. If I run into a colleague at the office, the cascade of activity of a slightly different group of neurons will allow me, in principle, to avoid confusing her with the president, and her office with the famous oval room. Distinct groups of neurons code for different faces and places—and because these neurons are tightly interconnected, the mere sight of the White House may evoke Trump's face, while I may have trouble recognizing my colleague out of context, for instance, if I run into her at the gym.

Let us now suppose that upon seeing the president in the Oval Office, my emotional systems judge this experience important enough to be stored in memory. How does my brain go about recording it? To cement the event, the neurons that were recently activated undergo major physical changes. They modify the strength of their interconnections, thus increasing the group support and making it more likely that this set of neurons will fire in the future. Some synapses become physically larger and may even get duplicated. The target neurons sometimes grow new spines, terminal boutons, or dendrites. All these anatomical modifications imply the expression of new genes, over the course of several hours or even days. These changes are the physical basis of learning: collectively, they form a substrate for memory.

Once a synaptic memory is formed, the neurons can now rest: when they stop firing, the memory remains dormant, unconscious but inscribed in the very anatomy of my neuronal circuits. In the future, thanks to those connections, an external clue (say, a photo of the presidential office) may suffice to produce a cascade of neuronal activity in the original circuit. This cascade will restore a pattern of neural discharges similar to the moment the memory was made, ultimately allowing me to recognize Donald Trump's face. According to this theory, each restored memory is a reconstruction: remembering is attempting to play back the very same neuronal firing pattern that occurred in the same brain circuits during a past experience.

Memory therefore cannot be ascribed to a single region of the brain—it is distributed in most, if not all, brain circuits, because each of them is capable of changing its synapses in response to a frequent pattern of neural activity. But not all circuits play the same role. Even if the terminology remains vague and continues to evolve, researchers distinguish between at least four kinds of memories.

- Working memory retains a mental representation in active form for a few seconds. It mainly relies on the vigorous firing of many neurons in the parietal and prefrontal cortices, which in turn support neurons in other, more peripheral regions.[9] Working memory is typically what allows us to keep a phone number in mind: during the time it takes us to type it into our smartphone, certain neurons support one another and thus keep the information in an active state. This type of memory is primarily based on the maintenance of a sustained pattern of activity—although it was recently discovered that it probably also involves short-lived synaptic changes,[10] allowing the neurons to go briefly dormant and quickly return to their active state. At any rate, working memory never lasts more than a few seconds: as soon as we get distracted by something else, the assembly of active neurons fades away. It is the brain's short-term buffer, keeping in mind only the hottest, most recent information.

- Episodic memory: The hippocampus, a structure located in the depths of the cerebral hemispheres below the cortex, records the unfolding episodes of our daily lives. Neurons in the hippocampus seem to memorize the context of each event: they encode where, when, how, and with whom things happened. They store each episode through synaptic changes, so we can remember it later. The famous patient H.M., whose hippocampi in both hemispheres had been obliterated by surgery, could no longer remember anything: he lived in an eternal present, unable to add the slightest new memory to his mental biography. Recent data suggest that the hippocampus is involved in all kinds of rapid learning. As long as the learned information is unique, whether it is a specific event or a new discovery worthy of interest, the neurons in the hippocampus assign it a specific firing sequence.[11]

- Semantic memory: Memories do not seem to stay in the hippocampus forever. At night, the brain plays them back and moves them to a new location within the cortex. There, they are transformed into permanent knowledge: our brain extracts the information present in the experiences we lived through, generalizes it, and integrates it into our vast library of knowledge of the world. After a few days, we can still remember the name of the president, without having the slightest memory of where or when we first heard it: from episodic, the memory has now become semantic. What was initially just a single episode was transformed into long-lasting knowledge, and its neural code moved from the hippocampus to the relevant cortical circuits.[12]

- Procedural memory: When we repeat the same activity over and over again (tying our shoes, reciting a poem, calculating, juggling, playing the violin, cycling...), neurons in the cortex and other subcortical circuits eventually modify themselves so that information flows better in the future. Neuronal firing becomes more efficient and reproducible, pruned of any parasitic activity, unfolding unerringly and as precisely as clockwork. This is procedural memory: the compact, unconscious recording of patterns of routine activity. Here, the hippocampus does not intervene: through practice, the memory gets stored in an implicit storage space, primarily involving a subcortical set of neural circuits called the "basal ganglia." This is why the patient H.M., even without any conscious, episodic, hippocampus-related memory, could still learn new procedures. The researchers even taught him to write backwards while looking at his hand in a mirror. Having no memory of the numerous times he had practiced this before, he was flabbergasted to find out how good he was at what he believed to be a completely new trick!

TRUE SYNAPSES AND FALSE MEMORIES

In the unforgettable movie *Eternal Sunshine of the Spotless Mind* (2004), French director Michel Gondry imagines a company that specializes in selectively erasing memories from people's brains. Wouldn't it be useful to delete the memories that poison our lives, such as those that cause post-traumatic stress in war veterans? Or, on the contrary, could we paint the illusory canvas of a false memory?

Neuroscientists' mastery of the circuits involved in memory is such that we are no longer that far away from Michel Gondry's dream. Both manipulations have already been performed in mice by the team of another Nobel Prize winner, professor Susumu Tonegawa. He first placed a mouse in a room and gave it minor electric shocks. The mouse then avoided the room where this unpleasant event took place, indicating that this episode was ingrained into its memory. Indeed, Tonegawa's colleagues managed to visualize it. Using a sophisticated two-photon microscope, they could track which neurons were active at each instant, and they saw that, in the hippocampus, different groups of neurons were activated for room A, which had been associated with the electric shock, and room B, where nothing had happened.

Then the researchers tested whether they could play around with those episodic memories. While the animal was physically located in room A, they again gave it small electric shocks, but this time they artificially activated the population of neurons that encoded room B. This artificial conditioning was effective: afterwards, when the mouse went back to room B, it became alarmed and froze in fear. The bad memory was now attached to room B, where nothing had ever happened.[13] Reactivating a meaningful group of neurons had sufficed to awaken a memory and tie it with new information.

Tonegawa's team then turned the bad memory into a good one. Could the traumatic memory be erased? Yes. By reactivating the same room B neurons when the mice were put in the presence of partners of the opposite sex—a guaranteed good moment—the researchers succeeded in erasing the association with the electric shock. The mice, far from avoiding the cursed room B, began to explore it frantically as if they were searching for the erotic partners that they remembered.[14]

Another group of researchers adopted a slightly different strategy: they reawakened the initial group of neurons while, at the same time, weakening the synapses that linked them. Again, in the days that followed, the mouse no longer showed the slightest memory of the initial trauma.[15]

In the same line of thought, French researcher Karim Benchenane succeeded in implanting a new memory in the mouse's brain during its sleep.[16] Whenever an animal falls asleep, neurons in its hippocampus spontaneously reactivate the memories of the previous day, especially the places where the animal went (we will return to this in more detail in Chapter 10). Taking advantage of this fact, Benchenane waited for the sleeping mouse's brain to reactivate the neurons associated with a particular place in its enclosure— and then gave the animal a small injection of dopamine, the neurotransmitter of pleasure. Lo and behold, as soon as the mouse woke up, it scurried as fast as it could toward this location! What was initially a neutral location had acquired, during the night, a very special place in memory, as addictive as the sweetness of Provence or the first place we fell in love.

Closer to us humans, some animal experiments have begun mimicking the effects of schooling on the brain. What happens when a monkey learns letters, numbers, or how to use tools?[17] Japanese researcher Atsushi Iriki showed that a monkey could learn to use a rake to collect pieces of food that were placed too far away to be grabbed by hand. After a few thousand tests, the animal became as quick as an experienced casino croupier: it took him only a few tenths of a second to rake in each food morsel, with a flick of the wrist. The monkey even figured out how to use a medium-size rake to pull a second, longer rake to him, in order to reach food placed at a much farther distance! This type of tool learning triggered a whole cascade of changes in the brain. Energy consumption increased in a specific area of the cortex, the anterior parietal region—the same area that humans use to control hand movements, write, grab an object, or use a hammer or pliers. New genes were expressed, synapses blossomed, dendritic and axonal trees multiplied—and all these additional connections resulted in a 23 percent increase in cortex thickness in this expert monkey. Whole bundles of connections also underwent dramatic alterations: axons coming from a distant region, at the junction with the temporal cortex, grew several millimeters and invaded a part of the anterior parietal region which previously had no connections to these neurons.

These examples illustrate the degree to which the effects of brain plasticity extends in time and space. Let's review the main points together. A set of neurons which codes for an event or concept that we wish to memorize is activated in our brain. How is this memory saved? In the beginning is the synapse, the microscopic point of contact between two neurons. Its strength is increased when the neurons it links are jointly activated in short succession—this is Hebb's famous rule: neurons that fire together, wire together. A synapse that gets stronger is like a factory that increases its productivity: it recruits more neurotransmitters on the presynaptic side and more receptor molecules on the postsynaptic side. It also increases in size to accommodate them.

As a neuron learns, its very shape changes too. A mushroom-like structure called a "dendritic spine" forms at the place on the dendrite where the synapse lands. If necessary, a second synapse emerges to double the first. Other synapses that land on the same neuron are also strengthened.[18]

Thus, when learning is prolonged, the very anatomy of the brain ends up changing. With recent advances in microscopy—in particular, the revolution brought on by two-photon microscopes, based on lasers and quantum physics—synaptic and axonal buttons can now be directly seen growing with each learning episode, just like a tree in springtime. When accumulated, the dendritic and axonal changes can be substantial, on the order of millimeters, and they begin to become detectable in humans through MRI. Learning to play music,[19] read,[20] juggle,[21] or even drive a taxi in a big city[22] results in detectable improvements in the thickness of the cortex and the strength of the connections that link cortical regions: the highways of the brain improve the more we use them.

Synapses are the epitome of learning, but not the only mechanism of change in the brain. When we learn, the explosion of new synapses often forces the neurons to also grow additional branches, both on axons and on dendrites. Far away from the synapse, the useful axons surround themselves with a sheath of insulation—myelin, similar to the adhesive tape that is wrapped around electrical wires to insulate them. The more an axon is used, the more layers this sheath develops, thus insulating it better and better, allowing it to transmit information at a higher speed.

Neurons are not even the sole cellular players of the learning game. As learning progresses, their whole environment also changes, including the surrounding glial cells, which feed and heal them, and even the vascular network of veins and arteries that provide them with oxygen, glucose, and nutrients. At this stage, an entire neural circuit and its support structure have changed.

Some researchers challenge the dogma that synapses are the indispensable actor of all learning. Recent data suggest that Purkinje cells, a special type of neuron in the cerebellum, can memorize time intervals, and that synapses play no role in this learning process: the mechanism seems to be purely internal to the cell.[23] It is quite possible that the dimension of time, which is a specialty of the cerebellum, is stored in memory using a different evolutionary trick, one which is not based on synapses. Each cerebellar neuron, all by itself, seems to be able to store several time intervals, perhaps through stable chemical changes in its DNA.

Another frontier of research consists of clarifying how such learning-induced changes, whether synaptic or not, can implement the most elaborate types of learning that the human brain is capable of, based on the "language of thought" and the fast recombination of existing concepts. As we have seen, conventional models of artificial neural networks provide a reasonably satisfying explanation for how millions of changing synapses allow us to learn to recognize a number, an object, or a face. However, there is no truly satisfactory model of how synaptic changes in neural networks underlie language acquisition or mathematical rules. Moving from the domain of synapses to the symbolic rules that we learn in math class remains a challenge today. Let us keep an open mind, because we are far from fully understanding all the biological codes by which our brain stores our memories.

NUTRITION AS A KEY ELEMENT OF LEARNING

What is certain is that when we learn, massive biological changes occur: not only do neurons undergo change in their scaffolding of dendrites and axons, but the surrounding glial cells do too. All these transformations take time. Each learning experience requires a cascade of biological changes, which can spread over several days. Many genes that specialize in plasticity must be expressed, so that the cells produce the necessary proteins and membranes to lay down new synapses, dendrites, and axons. This process absorbs a lot of energy: a young child's brain consumes up to 50 percent of the body's energy balance. Glucose, oxygen, vitamins, iron, iodine, fatty acids . . . a great variety of nutriments is essential for successful brain growth. The brain does not feed just on intellectual stimulation. To make and break a few million synapses per second, it requires a balanced diet, oxygenation, and physical exercise.[24]

A sad episode illustrates the extreme sensitivity of the developing brain to proper nutrition. In November 2003, children in Israel suddenly became afflicted by an unknown sickness.[25] Overnight, dozens of babies flooded into pediatric hospitals across the country. They presented severe neurological symptoms: lethargy, vomiting, vision impairments, and vigilance problems, sometimes leading to coma or, for two of them, to death. A race against time began: What was this new disease, and what caused its abrupt emergence?

The investigation eventually traced it back to nutrition. All the sick babies had been bottle-fed with the same soy-based milk powder. The analysis of its formula confirmed the worst of fears: according to the label, the milk should have contained 385 milligrams of thiamine, better known as vitamin B1. In reality, there was no trace of it. When contacted, the manufacturer admitted that he had altered the composition of the milk at the beginning of 2003: for economic reasons, he had stopped adding thiamine. This vitamin, however, is an essential nutrient for the brain. Even worse, the body does not store thiamine, so its absence from one's diet quickly leads to a serious deficiency.

Neurologists already knew that thiamine deficiency in adults causes a severe neurological disorder, Wernicke-Korsakoff syndrome, most often seen in heavy drinkers. In the acute phase, this deficiency induces Wernicke's encephalopathy, which can be fatal. Mental confusion, eye movement disorders, inability to coordinate movements, and deficient alertness, sometimes leading to coma and death . . . its symptoms resembled those of the babies in Israel in every way.

The ultimate proof came from therapeutic intervention. As soon as the essential vitamin B1 was restored to the children's diet, their condition improved in a few days, and they were able to return home. It is estimated that between six hundred and one thousand Israeli babies were deprived of thiamine for two to three weeks during the first months of their lives. The restoration of a balanced diet clearly saved them. However, years later, they exhibited major language impairments. Israeli psychologist Naama Friedmann tested about sixty of them when they were six or seven years old. The majority suffered from huge deficits in language comprehension and production. Their grammar was particularly abnormal—after reading or hearing a sentence, they had trouble figuring out who did what to whom. Even the simple task of naming a picture, like that of a sheep, was difficult for some of them. However, their conceptual processing seemed intact: they knew how to associate, for example, the image of a ball of wool with that of a sheep rather than a lion. And in all other respects, in particular with regard to intelligence (the famous IQ test), they appeared normal.

This sad story illustrates the limits of brain plasticity. Learning a language is obviously based on the immense plasticity of the infant's brain. Any baby is capable of learning any language of the world, from the tones in Chinese to the clicks in Bantu of South Africa, because its brain changes adequately in response to immersion in a particular community. However, this plasticity is neither infinite nor magical: it is a strictly material process that requires specific nutritional and energetic inputs, and even a few weeks of deprivation can lead to permanent deficits. And because the organization of the brain is highly modular, these deficits may be restricted to a specific cognitive domain, such as grammar or vocabulary. The pediatric literature is full of similar examples. I could have mentioned, for instance, fetal alcohol syndrome, which is caused by exposure of the fetus to alcohol ingested by the mother. Alcohol is a teratogen, a substance that causes embryonic malformations of the body and brain: it is a true poison for the developing nervous system, one that should clearly be avoided throughout pregnancy. For dendritic trees to thrive, the garden of the brain must be provided with all the nutrients it needs.

THE POWERS AND LIMITS OF SYNAPTIC PLASTICITY

In a well-fed brain, how far can plasticity go? Can it completely rewire the brain? Can brain anatomy dramatically change according to experience? The answer is no. Plasticity is an adjustment variable, fundamental for learning but restricted and confined by all kinds of genetic constraints that make us what we are: the conjunction of a fixed genome and unique experiences.

It is time I tell you more about Nico, the young artist whose art I introduced you to in the first chapter (see figure 1 in the color insert). Nico creates his splendid paintings using only a single brain hemisphere, his left. At the age of three years and seven months, he underwent a surgical procedure called a "hemispherectomy"—the quasi-complete removal of a hemisphere—to put an end to his devastating epilepsy.

Supported by his family, his doctors, and Harvard School of Education researcher Antonio Battro, Nico managed to attend elementary school in Buenos Aires, then went to high school in Madrid until he was eighteen. Nowadays, his oral and written language, memory, and spatial skills are all excellent. He even got his university diploma in IT. Above all, he has this remarkable talent for drawing and painting.

Is this a good case of brain plasticity at work? Undoubtedly, considering that Nico's left hemisphere has mastered many functions that, in a normal person, are traditionally associated with the right hemisphere. For instance, Nico manages to pay attention to the entirety of a picture and can copy the spatial arrangement of a drawing; he understands the irony and intonations of a conversation, and can guess the thoughts of the people he speaks to. If the same lesion occurred in an adult brain, these functions would likely be irremediably damaged.

Yet Nico's plasticity was demonstrably limited: it was channeled and largely confined to neuronal circuits, which are the same as those of all other children. When we scanned Nico with a whole battery of tests, we found that he had managed to fit all his learned talents into his intact left hemisphere without upheaving its usual organization. In fact, all the traditionally right-sided functions had landed at left-hemisphere locations symmetrical to their usual positions! For example, the cortical region which responds to faces and which is usually located in the right temporal lobe was now located in the left hemisphere in Nico—but in a very precise spot, exactly symmetrical to its usual site, a place often activated (weakly) by faces in normal children. Thus, while his brain had reorganized, it remained submitted to the strong constraints of a preexisting organization common to all humans. The great fiber bundles of connections which, from birth and even in utero, run through all babies' brains had forced his learning to remain within the narrow limits of a universal cortical map.

The powers and the limits of brain plasticity are never as obvious as when we consider visual abilities. Not surprisingly, Nico is hemianoptic, which means that his vision is split in two: a right half where he sees perfectly (in both eyes), and a left half where he is totally blind (again in both eyes). Whenever he gazes at something, the right part appears normal, while the left is invisible—he has to shift his eyes in order to see it. Indeed, due to the crossing of the visual pathways, inputs from the left side of the visual field, which would normally land in Nico's right hemisphere, now fall into a void and cannot be processed. Twenty years of visual experience have not allowed Nico's brain to compensate for this fundamental wiring problem. The plasticity of his visual connections was obviously too modest, and the development of this part of his brain froze too early in childhood to prevent him from going blind in his left visual field.

Now, let me tell you about another young patient: a ten-year-old girl we know only by her initials, A.H.[26] This child, like Nico, has only her left hemisphere, but unlike him, she suffered from an embryonic malformation that caused the development of her right hemisphere to completely stop before seven weeks of gestation. In other words, A. H. spent virtually her entire life without a right hemisphere. Did early plasticity radically change her brain? No, but it did manage to intervene a little bit more than it could for Nico. Unlike him, she is able to see some light, shape, and motion in her left visual field, the one that should have projected to her absent right hemisphere. Her vision there is far from perfect, but she does detect light and movement in a region close to the center of her vision. Brain imaging shows that her visual brain areas are partially remapped (see figure 11 in the color insert). In the back of her intact left hemisphere, within the occipital cortex, which houses vision, there is a perfectly normal map of the right part of the world—but also small abnormal patches that respond to the left part. It would seem that axons from half her retina, which normally should have been blind, were redirected to the other side of the brain. This is an extreme case of prenatal plasticity—and even so, the reorganization is only partial and quite insufficient to restore normal vision. In the visual system, genetic constraints dominate, and plasticity acts only within its narrow bounds.

Scientists were curious to see how far back these genetic limits could be pushed. One experiment is particularly famous, in which MIT neuroscientist Mriganka Sur succeeded in transforming the ferret auditory cortex into a visual cortex.[27] To do so, during a small surgical intervention on the ferret fetus, he severed the auditory circuits that normally travel from the cochlea to the brain stem, then reach a precise region of the auditory thalamus, and finally enter the auditory cortex. These ferrets inevitably ended up deaf—but then a curious reorientation occurred, and visual fibers began to invade this disconnected auditory circuit, as if they were replacing the missing auditory inputs. Lo and behold, an entire area of the cortex that should have been dedicated to audition now responded to vision. It contained a whole map of neurons sensitive to light and to oriented lines, like in any visual cortex. The synapses adapted themselves to this new configuration and began to encode the correlations between neurons that were originally destined for hearing but had been recycled into vision processors.

Should we conclude from these data that cerebral plasticity is "massive" and that experience is what "organizes the cortex," as the most ardent blank-slate defenders would proclaim?[28] That is not Sur's conclusion at all. On the contrary, he insists that this is a pathological situation, and that reorganization is far from perfect: in the auditory cortex, the visual maps are not as well differentiated as they should be. The visual cortex is genetically prepared to support vision. During normal development, each cortical region specializes very early on, under the influence of numerous developmental genes. Axons find their way along predetermined chemical pathways that trace out proto-maps in the developing brain. Only at the end of the road are they subjected to the growing influence of incoming neuronal activity and can then adapt to it. The neuronal tapestry is fixed, and only small but significant stitches can change.

It is also important to understand that when synapses change, even under the influence of neuronal activity, it is not necessarily the environment which is leaving an impression on the brain. Rather, the brain can use synaptic plasticity to *self*-organize: it first generates activity patterns purely from within, in the absence of any input from the environment, and uses those activity patterns, in combination with synaptic plasticity, to wire its circuits. In utero, even before they receive any sensory input, the brain, the muscles, and even the retina already exhibit spontaneous activity (this is why fetuses move in the womb). Neurons are excitable cells: they can fire off spontaneously, and their action potentials self-organize into massive waves that travel through brain tissue. Even in the womb, random waves of neuronal spikes flow through the fetus's retinas, and upon reaching the cortex, although they do not carry any visual information in the strict sense of the term, these waves help organize the cortical visual maps.[29] Thus, synaptic plasticity initially acts without requiring any interaction with the outside world. It is only during the third trimester of gestation that the line between nature and nurture gradually blurs as the brain, which is already well formed, begins to adjust to both inner and outer worlds.

Even after birth, random neuronal firing unrelated to sensory inputs continues to flow through the cortex. Very slowly, this endogenous activity evolves under the influence of the sensory organs. This process can be given a precise interpretation within the theoretical framework of the "Bayesian brain."[30] The initial endogenous activity represents what statisticians call the *prior*: the expectations of the brain, its evolutionary assumptions prior to any interaction with the environment. Later, these assumptions gradually adjust to environmental signals, so that after a few months of life, spontaneous neuronal activity resembles what statisticians call the *posterior*: the brain's probability distributions have changed to more and more closely reflect real-world statistics. During brain development, the internal models that we carry in our neuronal circuits get refined as each of them compiles statistics from its sensory inputs. The final result is a compromise, a selection of the best internal model from those that our prior organization makes available.

WHAT IS A SENSITIVE PERIOD?

We have just seen that brain plasticity is both vast and limited. All bundles of connections can and must change as we live, mature, and learn. However, the main ones are already in place from birth and remain essentially the same in all of us. Everything we learn results from small adjustments, mainly at the level of microcircuits, often on the scale of a few millimeters. As neurons mature and their terminal branches grow new synaptic buttons onto other neurons, the circuits they form remain firmly rooted within a limited genetic envelope. In response to the environment, neuronal pathways can change their local connectivity, their strength, and also their myelination, surrounding themselves with an insulating sheath of myelin, which accelerates their messages and thus facilitates the transmission of information from one region to another—yet they cannot reorient themselves at will.

This spatial constraint on long-distance connectivity is coupled with a temporal constraint: in many brain regions, plasticity is maximal only during a limited time interval, which is called a "sensitive period." It opens up in early childhood, peaks, and then gradually decreases as we age. The entire process takes several years and varies across brain regions: sensory areas reach their peak plasticity around the age of one or two years old, while higher-order regions such as the prefrontal cortex peak much later in childhood or even early adolescence. What is certain, however, is that as we age, plasticity decreases, and learning, while not completely frozen, becomes more and more difficult.[31]

The reason I affirm that babies are real learning machines is that, during their first years, their brains are the seat of an ebullient synaptic plasticity. The dendrites of their pyramidal neurons multiply at an impressive speed. At birth, the infant's cortex looks like a forest after a hurricane, sparsely populated by scattered, bare tree trunks. The first six months of life are literal springtime for the newborn brain, as neuronal connections and branches multiply until they form an inextricable jungle.[32]

Such a progressive complexification of neuronal trees could suggest that

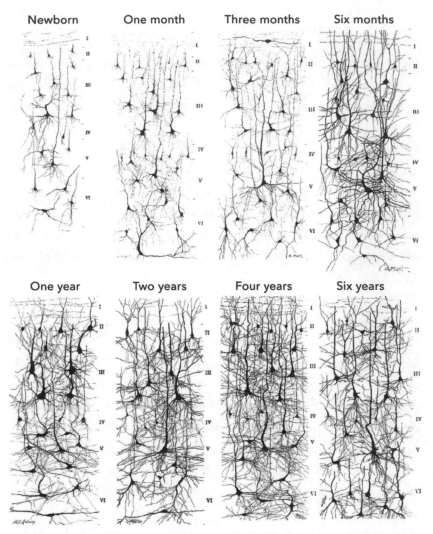

In the first two years of life, neuronal trees grow wildly until they form inextricable bushes. In the brain of a two-year-old child, the number of synapses is almost double that of an adult. In the course of development, dendritic trees are progressively trimmed under the influence of neuronal activity. Useful synapses are preserved and multiply, while unnecessary ones are eliminated.

the environment leaves its impression on the brain and forces it to grow as it stocks more and more data. The reality, however, is much more convoluted. In the immature brain, synapses do not emerge in direct proportion to how much learning occurs. Rather, they are created in excessive numbers, and the role of the environment is to keep them or prune them, depending on their utility for the organism. During early childhood, the density of synapses reaches twice that of an adult, and only then does it slowly decrease. In each region of the cortex, incessant waves of overproduction are followed by a selective retraction of useless synapses or, on the contrary, a multiplication of those synapses and dendritic and axonal branches that have proven their worth. Ponder this the next time you look at a young child: every second that goes by, several million synapses are created or eliminated in her brain. This effervescence largely explains the existence of sensitive periods. In early childhood, the whole dendritic and synaptic foliage is still highly malleable; the more the brain matures, the more learning is confined to marginal changes.

Remarkably, these waves of synaptic overproduction and pruning do not occur everywhere at the same time.[33] The primary visual cortex, like other sensory regions, matures much faster than higher-level cortical areas. The organizing principle seems to be to quickly stabilize the brain's inputs by freezing cortical organization in early sensory areas, while leaving the high-level areas open to change for much longer. Thus, regions higher up in the cortical hierarchy, such as the prefrontal cortex, are the last to stabilize: they continue to change during adolescence and beyond. In the human species, the peak of synaptic overproduction ends around two years of age in the visual cortex, three or four years of age in the auditory cortex, and between five and ten years of age in the prefrontal cortex.[34] Myelination, the wrapping of an insulator around axons, follows the same pattern.[35] In the first months of life, neurons in sensory areas are the first to benefit from an insulating sheet of myelin. As a result, visual information processing accelerates dramatically: the information transmission delay from the retina to the visual areas drops from a quarter to a tenth of a second in the first few weeks of life.[36] This insulation is much slower to reach the fiber bundles that project to the frontal cortex, the seat of abstract thought, attention, and planning. For years, young children possess a hybrid brain: their sensory and motor circuits are quite mature, while their higher-level areas continue to operate at the slow speed of unmyelinated circuitry. As a result, during the first year of their lives, it takes them up to four times longer than an adult to become aware of basic information, such as the presence of a face.[37]

In sync with those successive waves of synaptic overproduction and myelination, the sensitive periods for learning open and close at different times depending on the brain regions involved. Early sensory areas are among the first to lose their ability to learn. The best-studied example, in both humans and animals, is binocular vision.[38] To compute depth, the visual system merges the information from both of our eyes. Such "binocular fusion," however, happens only if the visual cortex receives high-quality inputs from both eyes during a well-defined sensitive period, which lasts a few months for cats and a few years for humans. If, during this period, one eye remains closed, or blurry, or misaligned because the child suffers from being severely cross-eyed, then the cortical circuit responsible for the fusion of the eyes fails to form, and the resulting loss is permanent. This condition, known as "amblyopia," or "lazy eye," must be corrected in the first years of life, ideally before three years of age—otherwise the wiring of the visual cortex remains forever impaired.

Another example of a sensitive period is the one that allows us to master the sounds of our native language. Babies are champions of learning languages: at birth, they distinguish all the phonemes of all possible languages. Wherever they are born and whatever their genetic background, all they have to do is immerse themselves in a language bath (which can be monolingual, bilingual, or even trilingual), and in a few months, their hearing becomes attuned to the phonology of the language that surrounds them. As adults, we have lost this remarkable learning ability: as we have seen, Japanese-speaking individuals can spend a lifetime in an English-speaking country without ever being able to distinguish the sound /R/ from the sound /L/, forever confusing "right" with "light," "red" with "led," and "election" with "erection." But, dear British or American reader, don't feel a sense of superiority, because, as a native English speaker, you will never be able to distinguish the dental and retroflex versions of the consonant /T/ that any Hindi speaker perceives as a no-brainer, nor the short and long vowels of Finnish or Japanese, nor the four kinds of tones of Chinese.

Research shows that we lose this ability toward the end of the first year of life.[39] As babies, we unconsciously compile statistics about what we hear, and our brain adjusts to the distribution of phonemes used by those around us. Around twelve months of age, this process converges and something freezes in our brain: we lose the ability to learn. Except in extraordinary circumstances, we will never again be able to pass ourselves off as native speakers of Japanese, Finnish, or Hindi—our phonology is (almost) set in stone. It takes immense effort for an adult to recover the ability to discriminate sounds in a foreign language. Only with intense and focused rehabilitation, first amplifying the differences between /R/ and /L/ to make them audible, then gradually reducing them, can a Japanese adult succeed in partially recovering the discrimination of these consonants.[40]

This is why scientists speak of a sensitive period rather than a critical period: the capacity for learning shrinks but never truly reaches zero. In adulthood, the residual ability to acquire foreign phonemes varies significantly across people. For most of us, trying to properly speak a foreign language in adulthood is an unfathomable endeavor—and this is why most French visitors to the United States sound like Inspector Clouseau in *The Pink Panther* ("Vere iz ze téléfawn?"). Remarkably, however, some people maintain a capacity to learn the phonology of foreign languages, and this competence can be partially predicted by the size, shape, and number of connections of their auditory cortex.[41] These lucky brains apparently stabilized a more flexible set of connections—but they are clearly the exception rather than the rule.

The sensitive period for mastering the phonology of a foreign language closes fast: as early as the first years of life, a child is already much less competent than a baby who is a few months old. Hierarchically, higher levels of language processing, such as grammar learning, remain open a little longer, but start closing around puberty. We know this from studies of children who arrive in a foreign country as migrants or adoptees: they may excel in their new language, but they often have a small foreign accent and occasional syntax errors that give away their true origin. This gap is barely detectable in children who entered the country at the age of three or four, but it increases massively in young people who immigrated in adolescence or adulthood.[42]

A recent paper collected data from millions of second-language learners on the internet and used them to model the average human language-learning curve.[43] The results suggest that grammatical learning abilities decline slowly during childhood and drop sharply around the age of seventeen. Because it takes time to learn, researchers recommend starting well before the age of ten.

Progressive loss of second-language fluency

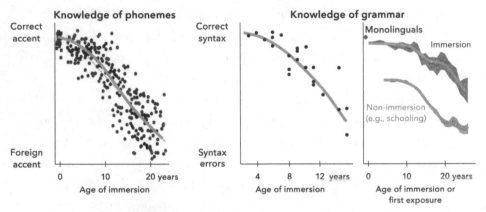

Reminiscence of the first language in adopted children

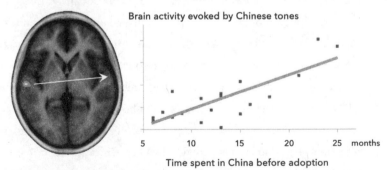

The capacity to acquire a foreign language decreases dramatically with age, suggesting the closure of a sensitive period for brain plasticity. The later you learn a language, the lower your chances of producing it without a foreign accent or grammatical errors (top). Conversely, the longer adopted children spend in their country of origin before leaving, the more their brains maintain a dormant, unconscious trace of their language of origin (bottom).

Furthermore, they emphasize the value of an immersive stay in the country of interest, because nothing beats social interaction: success is much better if you need to speak a foreign language in order to get served lunch or get on a bus than if you learn in a classroom or through watching TV. Once again, earlier is better: brain plasticity for grammar learning seems to drastically shrink at the end of puberty (although, not all of this drop may be imputed to a loss of brain plasticity; other factors related to motivation and socialization probably play a role too).

Up until now, we have considered only second-language acquisition, but note that this is an impure situation—this competence declines relatively slowly, over a decade or so, and never quite drops to zero, possibly because it relies, at least in part, on a brain already molded by the acquisition of a first language. What would happen if a child were deprived of all exposure to any language during the first years of life? Legend has it that Pharaoh Psamtik I was the first to ask this very question. He left two children in the care of a shepherd with strict prohibition against speaking to them—yet both babies eventually spoke . . . in Phrygian! This "experiment" was allegedly repeated by Emperor Frederick II in the thirteenth century, by James IV, king of Scotland, in the fifteenth century, and by Jalaluddin Muhammad Akbar, head of the Mughal Empire, in the sixteenth century—and some of these children, deprived of any language, supposedly died. (Lacanian psychoanalysts go crazy over this story.)

Alas, there is no need to spread such fables, for this situation occurs rather regularly in every country of the world: every day, children are born deaf, and if they are not helped, they remain prisoners in their bubble of silence. We now know that it is essential, as early as the first year of life, to give them a language: either sign language, which is the most natural (signed languages are real languages, and children who speak them develop quite normally), or a spoken language, when such children are able to receive a cochlear implant that partially restores their hearing. Here again, research has shown the necessity to act very quickly:[44] when children are implanted after the age of eight months, they already show permanent deficits in syntax. They never fully understand sentences where certain elements are moved around, a phenomenon called "syntactic movement." In the sentence "Show me the girl that the grandmother combs," it is not obvious that the first noun phrase, "the girl," is actually the object of the verb "combs" and not its subject. When deaf children receive a cochlear implant after the age of one or two years, they remain unable to understand such sentences, and they fail a test which requires choosing between a picture where the grandmother combs the girl's hair and another where the girl combs the grandmother's hair.

Early childhood seems to be an essential phase for the development of syntactic movement: toward the end of the first year of life, if the brain is deprived of any linguistic interactions, brain plasticity for this aspect of syntax closes. Remember the dying children in Israel in 2003: a few weeks of thiamine deprivation, in the first months of their lives, were enough to make them forever lose a sense for syntax. These results converge with other studies conducted on feral children who were abandoned by their families, such as the famous Victor of Aveyron (c. 1788–1828), and with research on abused children, such as the little American girl ironically named Genie and brought up (or rather brought down) in a closet for over thirteen years without being spoken to. Once Victor and Genie were brought back to civilization after so many years, they did start to speak and acquired some vocabulary, but their grammar remained permanently compromised.

Language learning thus provides an excellent example of sensitive periods in humans, both for phonology and for grammar. It is also a good illustration of the modularity of the brain: while the grammar and sounds of language freeze, other functions such as the capacity to learn new words and their meanings remain open throughout life. This residual plasticity is precisely what allows us to learn, at any age, the meanings of new words, such as *fax*, *iPad*, *meme*, and *geek*, or even humorous neologisms such as *askhole* (someone who keeps asking stupid, pointless questions) or *chairdrobe* (that pile of clothes we put on a chair instead of in a closet or dresser). For vocabulary acquisition, fortunately, our adult brain continues to exhibit a certain level of childlike plasticity throughout life—although the biological reason why lexical circuits do not suffer from a sensitive period is currently unknown.

A SYNAPSE MUST BE OPEN OR CLOSED

Why does synaptic plasticity close up? What biological mechanisms interrupt it? The origin of the opening and closing of sensitive periods is a major research topic in contemporary neuroscience.[45] The closing of the sensitive period seems to be related to the balance between excitation and inhibition. In children, excitatory neurons are rapidly effective, while inhibitory neurons develop more gradually. Some neurons, which contain a protein called "parvalbumin," progressively surround themselves with a hard matrix, a sort of lattice called a "perineuronal net" that becomes increasingly tight and eventually prevents synapses from growing or moving around. Entangled in this rigid net, neural circuits are no longer free to change. If we could release the neurons from this straitjacket, for example, by applying a pharmacological agent such as fluoxetine (better known as Prozac), synaptic plasticity might return. This is a huge source of hope for the treatment of stroke, where patients have to relearn their lost abilities using the preserved areas surrounding the brain lesion.

Other factors are also at play in closing a sensitive period. For example, there is a protein called "Lynx1," which, when present in a neuron, inhibits the massive effects of acetylcholine on synaptic plasticity. Acetylcholine, which normally signals events of interest and enhances synaptic plasticity, therefore loses its effect on adult circuits invaded by Lynx1. Some researchers have tried to restore plasticity by tampering either genetically with Lynx1 or pharmacologically with acetylcholine mechanisms—with some promising success in animals.

Another exciting possibility, perhaps more easily applicable to humans, consists of applying a current that depolarizes neurons and brings them closer to their firing threshold.[46] As a result, the excitable circuit becomes more easily activated and modifiable. This burgeoning therapy once again brings hope to patients, particularly those stuck in a deep depression: the application of a small electric current through the scalp is sometimes enough to put them back on the right path.

One may wonder why the nervous system persists in restraining its own

plasticity. After an initial stage of intense plasticity, there must be some evolutionary advantage in closing the sensitive period and avoiding further
changes to brain circuitry. Simulations of neural networks show that low-level
neurons, at the early stages of the visual hierarchy, quickly acquire simple and
reproducible receptive fields, such as contour detectors. It is likely that, beyond
the first few months of life, there is no further gain associated with continuing
to update them, since this type of detector is already nearly optimal. Our
brains might as well save the energy cost associated with the growth of synaptic and axonal buds. Moreover, changing the organization of early sensory
areas, the foundation on which all perception rests, risks creating havoc in
higher-level areas. From this perspective, after some time, it is probably worth-
while to leave these sensory neurons alone—and this is probably why evolution has settled on mechanisms that close off the sensitive period in sensory
areas at an earlier point in development than in higher-level associative areas.

The good side of things is that, because our circuits freeze up, we get
to keep, for our entire lives, a stable, unconscious synaptic trace of what we
learned as children. Even if those early acquisitions are later rendered obso-
lete, for example, because they are overridden by more recently acquired knowledge, our brain circuits retain a dormant trace of our beginnings. A remarkable
example is the case of children adopted after infancy who have to learn a
second maternal language. In the latter half of the twentieth century, Korea
was one of the countries that massively resorted to international adoption.
Since 1958, over a period of forty years, nearly 180,000 Korean children were
adopted, and the vast majority (about 130,000) left for a far-away country,
with more than 10,000 reaching France. In our Paris-based research center,
Christophe Pallier and I scanned twenty of them as adults. Having arrived in
France between the ages of five and nine years old, these young men and
women had virtually no conscious recollection of their native land (except for
a few olfactory memories, especially for the smell of food!). Our scans showed
that their brains behaved essentially like that of a child born in France:[47] their
language areas, in the left hemisphere, responded strongly to French sentences, but not at all to Korean sentences (in any case, no more than to any

other unknown language, such as Japanese). At the lexical and syntactic level, therefore, it seemed that the new language had supplanted the old one.

And yet . . . with a more subtle approach, another group of researchers found that adopted children still harbor, deep in their cortex, a dormant trace of the sound patterns of their original language.[48] They scanned children between the ages of nine and seventeen who had spent only one year in China before they were adopted in Canada. And instead of simply letting them hear sentences, the researchers gave them the difficult task of discriminating the tonal patterns of Chinese. Brain imaging showed that, while native Canadian adults without any exposure to Chinese failed to hear these tones as language, merely processing them as a melody in the right hemisphere, the Chinese-Canadian adoptees, just like native Chinese, processed them as language sounds in a phonological region of the left hemisphere called the "planum temporale." Thus, this circuit seems to become engraved with a native language in the first year of life, and never fully reverses afterwards.

This is not the only example. I already explained how a child's lazy eye can forever affect the visual circuits in the brain if the problem is left unattended. Ethologist and neurophysiologist Eric Knudsen studied an animal model of this sensitive period effect. He raised young owls and had them wear prism glasses that shifted the entire visual field about twenty degrees to the right. With his glasses-wearing owls, he carried out the finest studies of the neural mechanisms of the sensitive period.[49] Only those owls that wore prisms during their youth were able to adjust to this unusual sensory input: their auditory responses shifted to align with the retina, thus enabling them to hunt based on synchronized hearing and night vision signals. Older owls, however, even after having worn prisms for weeks, failed miserably. Most interestingly, the animals trained during their youth harbored, for the rest of their lives, a permanent neuronal trace of their early experience. After learning, a two-route circuit was observed: some axons of the auditory neurons in the lower colliculus retained their normal position, while others reoriented to align with the visual map. When the prisms were removed, the owls quickly learned to reorient correctly; and as soon as the glasses were put back on, the animals immediately readjusted by shifting the auditory scene by twenty degrees. Like a *parfait* bilingual, they managed to switch from one *langue* to the other. Their brains kept a permanent record of the two sets of parameters and allowed them to change configuration in a heartbeat—just as the Chinese adoptees in Canada kept a cerebral trace of the sounds of their original language.

In our species too, early learning—be it from practicing the piano,

Early experience can profoundly shape our brain circuits. An owl can adjust to wearing glass prisms that shift its vision—but only when this abnormal experience takes place during its youth. The owl's auditory neurons, which locate objects by relying on the tiny delay between sounds reaching the right and left ears, adjust in order to align themselves with visual signals. Axons can be displaced by about half a millimeter. Following this early experience, the two circuits—normal and displaced—remain present throughout the owl's life.

developing binocular vision, or even acquiring our first words—leaves a permanent mark. As adults, we are faster to recognize the words that we first heard during our childhood, such as "bottle," "dad," or "diaper"—early synaptic plasticity has forever engraved them in our memory.[50] The juvenile cortex learns languages almost effortlessly and stores this knowledge in the permanent geometry of its axons and dendrites.

A MIRACLE IN BUCHAREST

The evidence of heightened brain plasticity in the first few years means that investing in early education should be a priority. Early childhood is a highly sensitive period when many of a child's brain circuits are most easily transformed. Later, the gradual loss of synaptic plasticity makes learning increasingly difficult—but let us not forget that this progressive freezing of neural circuits is precisely what enables our brain to keep a stable trace of all that we learned in our childhood. Those permanent synaptic marks eventually define who we are.

Although learning is easier when it occurs earlier, it would be profoundly wrong to heed the credo of the United States' "zero-to-three movement" and conclude that everything hinges on this sensitive period. No, most learning does not happen before three years. Fortunately for us, our brains remain pliable for many more years. After the blessed period of early childhood, neural plasticity diminishes, but it never disappears. It slowly weakens over time, starting with the peripheral sensory areas, but high-level cortical areas keep their potential for adaptation throughout our lives. This is why many adults successfully learn to play an instrument or speak a second language in their fifties or sixties. And this is also why educational interventions sometimes work miracles, especially when they are fast and intense. Rehabilitation may not restore all the subtleties of syntactic movement or the perception of Chinese tones, but it will succeed in transforming an at-risk child into a fulfilled and responsible young adult.

The Bucharest orphans provide a heartbreaking example of this remarkable resilience of the developing brain. In December 1989, Romania suddenly rose up against the communist regime. In less than a week, the citizens in revolt drove the dictator Nicolae Ceaușescu (1918–1989) and his wife out of power—both were hastily tried, convicted, and shot on Christmas Day. Shortly after, the world was appalled upon discovering the dreadful living conditions of the residents in this small corner of Europe. One of the most unbearable sights was that of the young, dead-eyed, and emaciated children abandoned across nearly six hundred Romanian orphanages. In those true death houses, close to 150,000 children were packed in and left almost entirely to themselves. The Ceaușescu regime was so profoundly convinced that the strength of a country lies in its youth that it had put into effect a delirious pro-birth policy. Everything was done to ensure births by the thousands, from massive taxation of single people and couples without children to the prohibition of contraception and abortion, and even the death penalty for those who chose to abort.... Couples who could not provide for their children had no choice but to hand them over to the state services. Hence the hundreds of orphanages, which, quickly overwhelmed, failed to provide hygiene, food, heating, and the minimum of human contact and cognitive stimulation essential for normal child development. This disastrous policy produced thousands of neglected children with major cognitive and emotional deficits in all areas.

After the country opened up its borders, several NGOs looked into this catastrophe. From this was born a very special research project, the Bucharest Early Intervention Project.[51] With the agreement of the Romanian state secretariat for child welfare, Harvard researcher Charles Nelson decided to study with scientific rigor the consequences of having lived in an orphanage, and the possibility of saving these children by placing them in foster families. As there was no proper placement program in Romania, he set up his own recruitment system and managed to find 56 volunteer families who were willing to adopt 1 or 2 orphans each. Yet this was a mere drop of water in the dark oceans of the Romanian orphanages: only 68 children were able to leave. Nelson's *Science* publication describes in detail the dramatic Dickensian moment when 136 children were gathered and numbered from 1 to 136, and then those numbers were drawn

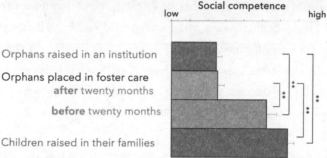

Childhood injuries leave their mark on the brain, but early intervention can minimize them. In Romanian orphanages during the Ceauşescu dictatorship, children were mistreated and deprived of interaction with adults. By the age of eight, most of these orphans showed major deficits in social skills, whether they had remained in an institution or been placed in a family after the age of twenty months. However, those who had been placed in foster care before the age of twenty months exhibited essentially normal skills.

at random from a large hat, determining who would remain in an orphanage and who would finally find a family home. This procedure may sound shocking, but what more could have been done? Because human resources were limited, a random draw was probably the fairest solution. In addition, the team continued to raise funds to lift more and more children out of their misery, as well as to advise the new Romanian government on its handling of institutionalized children, and a second *Science* article found that the initial study therefore met the ethical criteria for scientific research.[52]

The random draw, however, allowed a rigorous question to be asked: All other things being equal, did early placement in a foster family put these children back on their feet? The answer was positive, but highly age-dependent: only the children who were placed in a home before twenty months of age ended up much better off than those who had remained in the orphanage.

Dozens of previous studies had documented the dramatic effects of emotional and social isolation on brain development, and the Bucharest study was no exception: compared to children born in a typical family, all the orphans presented with severe deficits in cognitive function. Even fundamental aspects of brain functioning, such as glucose metabolism and total volume of gray matter, were deficient. After foster care, however, some of these measures sharply rose. Six years later, by the age of eight, the children who had been placed in a home before twenty months of age had made significant progress relative to the control group, to such an extent that they no longer differed from children raised in their families from birth. Several measures had normalized, including the strength of their brains' alpha waves, which is a marker of attention and vigilance. Social skills and vocabulary were also markedly improved.

This dramatic progress should not obscure the fact that these children continued to lag behind in other measures, including a persistent, and probably permanent, lack of gray matter. Most crucially, the children who had been adopted after twenty months exhibited severe impairments in all domains. Thus, no amount of family support can fully replace twenty months of lost love (and simple nutrition), and these children will always bear, in their brains, the scars of the serious deprivation they suffered. But the Bucharest orphans, like the adoptees from Korea, should remind us that we should never lose hope. Brain plasticity is certainly stronger in the young, but it remains present at any age. Early trauma may have a severe impact, yet the resilience of neural circuits is equally remarkable. Provided they are addressed as early as possible, many brain injuries are far from irreversible.

CHAPTER 6

Recycle Your Brain

LET'S SUMMARIZE WHAT WE HAVE COVERED SO FAR. THE BLANK-SLATE
assumption is clearly wrong: babies are born with considerable core knowledge, a rich set of universal assumptions about the environment that they will
later encounter. Their brain circuits are well organized at birth and give them
strong intuitions in all sorts of domains: objects, people, time, space, numbers....
Their statistical skills are remarkable—they already act as budding scientists,
and their sophisticated learning abilities allow them to progressively converge onto the most appropriate models of the world.

At birth, all the large fiber bundles of the brain are already in place. Brain
plasticity can, however, reorganize their terminal connections. Millions of
synapses undergo plastic changes every time we acquire new knowledge.
Enriching children's environments, for instance, by sending them to school,
can deeply enhance their brains and augment them with skills that they will
keep throughout their lives. This plasticity is not unconstrained, however. It
is restricted in space (on the order of a few millimeters), as well as in time—
many circuits begin to close off after a few months or years.

In this chapter, I look at the role that formal education plays in early brain
development. Education, indeed, raises a paradox: Why is it that *Homo sapiens*
can take chalk or a keyboard and start writing or making calculations? How

could the human species expand its capabilities in novel directions that previously played no part in its genetic evolution? That the human primate manages
to learn to read or to calculate should never cease to amaze us. As Vladimir
Nabokov (1899–1977) put it so well, "We are absurdly accustomed to the miracle of a few written signs being able to contain immortal imagery, involutions
of thought, new worlds with live people, speaking, weeping, laughing. What
if we awake one day, all of us, and find ourselves utterly unable to read?"[1]

I have studied at length the minds and brains of illiterate adults, whether
it be in Portugal, Brazil, or the Amazon—people who never had the chance
to go to school, simply because their families could not afford it or because
there were no schools nearby. Their skills are, in certain ways, profoundly
different:[2] not only are they incapable of recognizing letters, but they also
have difficulties recognizing shapes and distinguishing mirror images,[3] paying attention to a part of a face,[4] and memorizing and distinguishing spoken
words.[5] So much for Plato, who naively believed that learning to read would
ruin our internal memory by forcing us to rely on the external memory of
books. Nothing could be further from the truth. The myth of the illiterate
bard who effortlessly musters immense powers of memory is just that: a myth.
We all need to exercise our memory—and it gets better, not worse, by having
gone to school and learned to read.

The impact of education is even more striking in mathematics.[6] We discovered this by studying the many Amazon Indians who never had the chance
to go to school. First of all, many of them do not know how to precisely count
a collection of items. Many of their languages do not even include a counting system—they either have just a handful of words for "few" versus "many"
(like the Pirahã), or just fuzzy words for the numbers one to five (like the
Munduruku), and if they learn to count at all, for instance, using Spanish or
Portuguese number words, it is with a huge delay (like the Tsimane) compared to Western children.[7] Second, they possess only the rudiments of mathematical intuition: they distinguish basic geometrical shapes, understand the
organization of space, can navigate in a straight line, perceive the differences
between quantities such as thirty and fifty, and know that numbers can be

ordered from left to right. We inherit these skills from our evolution and share them with other animals as diverse as ravens, macaque monkeys, and freshly hatched chicks. However, education vastly increases these initial skills. For instance, uneducated Amazon Indians do not seem to understand that there is the same interval of +1 between any two consecutive numbers. Education massively overturns our sense of the number line: as we learn to count and to perform exact arithmetic, we discover that every number n has a successor $n + 1$. Eventually, we understand that all consecutive numbers are equidistant and form a linear scale—whereas very young children and unschooled adults consider this line to be compressed, since large numbers seem to be closer to each other than small ones.[8] If we had only an approximate sense of numbers like other animals do, we would be unable to distinguish eleven from twelve. We owe the refined precision of our number sense to education—and on this symbolic foundation rests the whole field of mathematics.

THE NEURONAL RECYCLING HYPOTHESIS

How does education revolutionize our mental skills, transforming us into primate readers of Nabokov, Steinbeck, Einstein, or Grothendieck? As we have seen, all that we learn passes through the modification of pre-established brain circuits, which are largely organized at birth but remain capable of changing on the scale of a few millimeters. Thus, all the diversity of human culture must fit within the constraints imposed by our neuronal nature.

To resolve this paradox, I have formulated the neuronal recycling hypothesis.[9] The idea is simple: while synaptic plasticity makes the brain malleable—especially in humans, where childhood lasts for fifteen or twenty years—our brain circuits remain subject to strong anatomical constraints, inherited from our evolution. Therefore, each new cultural object we invent, such as the alphabet or Arabic numerals, must find its "neuronal niche" in the brain: a set of circuits whose initial function is sufficiently similar to its new cultural role, but also flexible enough to be converted to this new use. Any cultural learning must rely on the repurposing of a preexisting neural architecture, whose properties it recycles. Education must therefore fit within the inherent limits of our neural circuits, by taking advantage of their diversity, as well as of the extended period of neural plasticity which is characteristic of our species.

According to this hypothesis, to educate oneself is to recycle one's existing brain circuits. Over the millennia, we have learned to make something new out of something old. Everything we learn at school reorients a preexisting neural circuit in a new direction. To read or calculate, children repurpose existing circuits that originally evolved for another use, but which, due to their plasticity, manage to adapt to a new cultural function.

Why did I coin this strange term, "neuronal recycling"? Because the corresponding French word, *recyclage*, perfectly combines two ideas that characterize what happens in our brain—a reuse of some material with unique properties, and also a reorientation toward a new career:

- Recycling a material means giving it a second life by reintroducing it into a novel production cycle. Such reuse of materials, however, is limited: one cannot build a car out of recycled paper! Each material possesses intrinsic qualities that make it more or less suitable for other uses. Similarly, each region of the cortex—by virtue of its molecular properties, local circuits, and long-range connections—possesses its own characteristics from birth on. Learning must conform to these material constraints.

- In French, the term recyclage also applies to a person who is training for a new job: it means to receive additional training in order to adapt to an unexpected change in one's career. This is exactly what happens to our cortex when we learn to read or to do math. Education grants our cortex new functions that go beyond the normal abilities of the primate brain.

With neuronal recycling, I wanted to distinguish the fast learning of a new cultural skill from the many other situations where biology, in the course of a slow evolutionary process, makes something new with something old. Indeed, in the Darwinian process of evolution by natural selection, repurposing of older materials is common. Genetic recombination can spruce up ancient organs and turn them into elegant, innovative machines. Bird feathers? Old thermal regulators converted into aerodynamic flaps. Reptilian and mammalian legs? Antediluvian fins. Evolution is a great tinkerer, says the French Nobel Prize–winning biologist François Jacob (1920–2013): in its workshop, lungs convert into floating organs, an old piece of the reptilian jaw becomes the inner ear, and even the sneer of hungry carnivores turns into Mona Lisa's delicate smile.

The brain is no exception. Language circuits, for instance, may have appeared during hominization through the duplication and repurposing of previously established cortical circuits.[10] But such slow genetic modifications do not fall under my definition of neuronal recycling. The appropriate term is "exaptation," a neologism coined by Harvard evolutionist Stephen Jay Gould (1941–2002) and Yale paleontologist Elisabeth Vrba and based on the word "adaptation." An old mechanism is exapted when it acquires a different use in the course of Darwinian evolution. (A simple mnemonic may help: exaptation makes your ex apt to a new task!) Because it is based on the spreading of genes through a population; at the species level, exaptation acts over tens of thousands of years. Neuronal recycling, on the other hand, occurs within an individual brain and on a much shorter time frame, anywhere from days to years. Recycling a brain circuit means reorienting its function without genetic modification, merely through learning and education.

My intent in formulating the neuronal recycling hypothesis was to explain the particular talent of our species for going beyond its usual ecological niche. Humans, indeed, are unique in their ability to acquire new skills, such as reading, writing, counting, doing math, singing, dressing, riding a horse, and driving a car. Our extended brain plasticity, combined with novel symbolic learning algorithms, has given us a remarkable ability for adaptation—and our societies have discovered means of further amplifying our skills by subjecting children, day after day, to the powerful regime of school.

To emphasize the singularity of the human species is not to deny, of course, that neuronal recycling, on a smaller scale, also exists in other animals. Recent technologies have made it possible to record the activity of the same hundred neurons for several weeks, while monkeys acquire a new skill—and, thus, to put the recycling view to a strong test. These experiments were able to address a simple but profound prediction of the theory: Can learning ever radically change the neural code in a given brain circuit, or, as the recycling view would predict, does learning solely repurpose the circuit?

In a recent experiment, using a brain-computer interface, researchers asked a monkey to learn to control its own brain. They taught the animal that to make a cursor go right, it had to activate ten specific neurons; and to make the cursor go up, it should activate ten other cells; and so on.[11] Remarkably, this procedure worked: in a few weeks, the animal learned to bend the activity of ten arbitrarily chosen neurons in order to make the cursor move at will. However—and this is the key—the monkey was able to get the cursor to move only if the neuronal activity that it was asked to produce did not deviate too much from what its cortex was already spontaneously producing before training. In other words, what the monkey was asked to learn had to fit within the repertoire of the neuronal circuit that it was asked to retrain.

To appreciate what the researchers showed, it is important to realize that the dynamics of brain circuits are constrained. The brain does not explore every configuration of activity that it might be able to access. In theory, in a group of a hundred neurons, activity could span a hundred-dimension space, yielding an unfathomable number of states (if we consider that each neuron could be on or off, this number exceeds 2^{100}, or over a thousand billion billion billion). Yet, in reality, brain activity visits only a fraction of this humongous universe, typically restricted to around ten dimensions. With this idea in mind, the constraint on learning can be formulated succinctly: a monkey can learn a novel task only if what we ask of its cortex "fits" within this preexisting space. If, on the other hand, we ask the monkey to activate a combination of neurons that is never observed in prior activity, it fails dramatically.

Note that the learned behavior itself may be radically new—who could have foreseen that a primate would one day control a cursor on a computer screen? However, the neuronal states that make this behavior possible must fit within the space of available cortical activity patterns. This result directly validates a key prediction of the neuronal recycling hypothesis—the acquisition of a novel skill does not require a radical rewriting of cortical circuits as if they were a blank slate, but merely a repurposing of their existing organization.

It is becoming increasingly clear that each region of the brain imposes its own set of constraints on learning. In a region of the parietal cortex, neural activity is generally confined to a single dimension, a straight line in high-dimensional space.[12] These parietal neurons encode all incoming data on an axis ranging from small to large—they are therefore ideally suited to encode quantities and their relative sizes. Their neural dynamics may seem extraordinarily limited, but what seems like a handicap could actually be an advantage when it comes to representing quantities, such as size, number, area, or any other parameter that can be ordered from small to large. In a sense, this part of the cortex may be pre-wired to encode quantities—indeed, it is systematically recruited as soon as we manipulate quantities along a linear axis, from numbers to social status (who is "above" whom on the social ladder).[13]

For another example, consider the entorhinal cortex, a region of the temporal cortex that contains the famous grid cells that map out space (which I described in Chapter 4). In this region, the neural code is two-dimensional: even if there are millions of neurons in this part of the brain, their activity cannot help but remain confined to a plane, or, technically, a two-dimensional manifold in high-dimensional space.[14] Again, this property, far from being a drawback, is obviously perfectly suited to form a map of the environment, as seen from above—and in fact, we know that this region hosts the mental GPS by which a rat locates itself in space. Remarkably, recent work has shown that this same region also lights up as soon as we have to learn to represent any data on a two-dimensional map, even if these data are not directly spatial.[15] In one experiment, for instance, birds could vary in two dimensions: the length of their neck, and the length of their legs. Once the human participants had learned to represent this unusual "bird space," they used their entorhinal cortex, along with a few other areas, to navigate it mentally.

And the list could go on and on: the ventral visual cortex excels at representing visual lines and shapes, Broca's area codes for syntactic trees,[16] and so forth. Each region has its own preferred dynamics to which it remains faithful. Each projects its own space of hypotheses onto the world: one tries to fit the incoming data on a straight line, another tries to display them on a map, a third on a tree.... These hypotheses spaces precede learning and, in a certain way, make it possible. We can, of course, learn new facts, but they need to find their neuronal niche, a representation space adapted to their natural organization.

Let us now see how this idea applies to the most fundamental areas of school learning: arithmetic and reading.

MATHEMATICS RECYCLES THE CIRCUITS FOR APPROXIMATE NUMBER

Let us first take the example of mathematics. As I explained in my book *The Number Sense*,[17] there is now considerable evidence to show that math education (like so many other aspects of learning) does not get imprinted onto the brain like a stamp on melted wax. On the contrary, mathematics molds itself into a preexisting, innate representation of numerical quantities, which it then extends and refines.

In both humans and monkeys, the parietal and prefrontal lobes contain a neural circuit that represents numbers in an approximate manner. Before any formal education, this circuit already includes neurons sensitive to the approximate number of objects in a concrete set.[18] What does learning do? In animals trained to compare quantities, the amount of number-detecting neurons grows in the frontal lobe.[19] Most important, when they learn to rely on the numerical symbols of Arabic digits, rather than on the mere perception of approximate sets, a fraction of these neurons become selective to such digits.[20] This (partial) transformation of a circuit in order to incorporate the cultural invention of numerical symbols is a great example of neuronal recycling.

In humans, when we learn to perform basic arithmetic (addition and subtraction), we continue to recycle that region, but also the nearby circuitry of the posterior parietal lobe. That region is used to shift our gaze and our attention—and it seems that we reuse those skills to move in number space: adding activates the same circuits that move your attention to the right, in the direction of larger numbers, while subtracting excites circuits that shift your attention to the left.[21] We all possess a kind of number line in our heads, a mental map of the number axis on which we have learned to accurately move when we perform calculations.

Recently, my research team has provided a more stringent test of the recycling hypothesis. With Marie Amalric, a young mathematician turned cognitive scientist, we wondered whether the same circuits of the parietal lobe continue to be used to represent the most abstract concepts in mathematics.[22] We recruited fifteen professional mathematicians and scanned their brains with functional MRI while we presented them with abstruse mathematical expressions that only they could understand, including formulas like $\int_s \nabla \times F \cdot dS$ and statements such as "Any square matrix is equivalent to a permutation matrix." As we predicted, these high-level mathematical objects activated the very same brain network that activates when a baby sees one, two, or three objects,[23] or when a child learns to count (see figure 12 in the color insert).[24] All mathematical objects, from Grothendieck's topoi to complex manifolds, or functional spaces, find their ultimate roots in the recombination of elementary neural circuits present during childhood. All of us, at any stage of the cultural construction of mathematics, from elementary school students to Fields Medal winners, continually refine the neural code of that specific brain circuit.

And the organization of that circuit is under strong hereditary constraints, those of the universal genetic endowment that makes us human. While learning allows it to accommodate many new concepts, its overall architecture remains the same in all of us, independent of experience. My colleagues and I obtained strong support for this assertion when we studied the brain organization of mathematicians whose sensory experience, since childhood, has been radically different: blind mathematicians.[25] However surprising this may seem, it is not uncommon for a blind person to become an excellent mathematician. Perhaps the best-known blind mathematician is Nicholas Saunderson (1682–1739), who became blind around the age of eight and was so brilliant that he ended up holding the chair of Isaac Newton at Cambridge University.

Saunderson is no longer available for a brain scan, but Marie Amalric and I managed to contact three contemporary blind mathematicians, all of whom held university positions in France. One of them, Emmanuel Giroux, is a true giant of mathematics and currently heads a laboratory of sixty people at the École normale supérieure in Lyon. Blind since the age of eleven, he is most well-known for his beautiful proof of an important theorem of contact geometry.

The very existence of blind mathematicians refutes Alan Turing's empiricist view of the brain as a "notebook" with "lots of blank sheets" that sensory experience progressively fills out. Indeed, how could blind people infer, from such a distinct and restricted experience, the very same abstract notions as sighted mathematicians if they did not already possess the circuits capable of generating them? As Emmanuel Giroux says, paraphrasing *The Little Prince*, "In geometry, what is essential is invisible to the eye. It is only with the mind that you can see well." In mathematics, sensory experiences do not matter much; it is the ideas and concepts that do the heavy lifting.

If experience determined the organization of the cortex, then our blind mathematicians, who learned about the world from touch and hearing, would activate, when they do mathematics, brain areas very different from those of the sighted. The neuronal recycling hypothesis, on the contrary, predicts that the neural circuits of mathematics should be fixed—only a specific set of brain areas, present at birth, should be capable of repurposing themselves to host such ideas. And this is indeed exactly what we found when we scanned our three blind professors. As we expected, when they visualized a mathematical statement and assessed its truth value, they recruited the very same parietal and frontal lobe pathways as a sighted mathematician (see figure 13 in the color insert). Sensory experiences were irrelevant: only this circuit could accommodate mathematical representations.

The only difference is that, when our three blind mathematicians thought about their favorite field, they also recruited an additional region of the brain: their early visual cortex, in the occipital pole, the brain region that, in any sighted person, processes the images that impinge on the retina! In fact, this is a result that Cédric Villani, another brilliant mathematician and Fields Medal winner, had intuitively predicted. When we discussed this experiment prior to running it, he jokingly said to me, "You know, Emmanuel Giroux is a truly great mathematician, but he is also very fortunate: because he is blind, he can devote even more cortex to math!"

Villani was right. In people with normal eyesight, the occipital region is too busy with early vision to perform any other function, such as mathematics. In the blind, however, it is released from this visual role, and instead of remaining inactive, it transforms itself to perform more abstract tasks, including mental calculation and mathematics.[26] And in people who are born blind, this reorganization seems to be even more extreme: the visual cortex exhibits totally unexpected responses, not only to numbers and math, but also to the grammar of spoken language, similar to Broca's area.[27]

The reason for such abstract responses in the visual cortex of blind people remains the subject of theoretical debate: Does this total reorganization of the cortex represent a genuine case of neuronal recycling, or is it merely an extreme proof of brain plasticity?[28] In my opinion, the scales are tipped in favor of the neuronal recycling hypothesis, because there is evidence that the preexisting organization of this region is not erased, as it would be if brain plasticity acted as a sponge capable of wiping clean the visual cortex's chalkboard. Indeed, the visual cortex of the blind seems to largely maintain its normal connectivity and neural maps[29] while reusing them for other cognitive functions. In fact, because this part of the cortex is very large, one finds "visual" regions in the brains of blind people that respond not only to math and language, but also to letters and numbers (presented in Braille), objects, places, and animals.[30] Most remarkable, in spite of such radical differences in sensory experience, these category-selective areas tend to be located at the same place in the cortex of sighted and blind individuals. For instance, the region of the brain that responds to written words is located at exactly the same

place in a blind person as it is in a sighted reader—the only difference is that
it responds to Braille rather than to printed letters. Once again, the function
of this region seems to be largely determined by its genetically controlled
connections to language areas, in addition, perhaps, to other innate properties,
and therefore does not change when sensory inputs do.[31] The blind entertain
the very same categories, ideas, and concepts as sighted people—using very
similar brain regions.

The neuronal recycling view of mathematics is not supported solely by
the fact that the most elementary concepts $(1 + 1 = 2)$ and the most advanced
mathematical ideas $(e^{-i\pi} + 1 = 0)$ make use of the same brain regions. Other
discoveries, of a purely psychological nature, indicate that the mathematics
we learn in school is based on the recycling of old circuits devoted to approximate quantities.

Think of the number five. Right now, your brain is reactivating a representation of an approximate quantity close to four and six and far from one
and nine—you are activating number neurons very similar to those found in
other primates, with a tuning curve that peaks around five, but also with
weights in the nearby quantities four and six. The fuzzy tuning curve of those
neurons is the main reason why it is hard, at a glance, to know if a set of objects
contains exactly four, five, or six items. Now, please decide if five is larger or
smaller than six. It seems instantaneous—you get to the correct answer
(smaller) in an instant—and yet, experiments actually show that your answer
is influenced by the approximate quantities: you are much slower when the
numbers are close, like five and six, than when they are further apart, like five
and nine, and you also make more errors. This distance effect[32] is one of the
signatures of an ancient representation of numbers that you recycled when
you learned to count and calculate. No matter how much you try focusing on
the symbols themselves, your brain can't help but activate the neural representations of these two quantities, which overlap more the closer they are.
Although you are trying to think of "exactly five," using all the symbolic
knowledge that you acquired at school, your behavior betrays the fact that this

knowledge recycles an evolutionarily older representation of approximate quantity. Even when you simply have to decide whether two numbers such as eight and nine are the same or different, which should be immediate, you continue to be influenced by the distance between them—and, interestingly, exactly the same finding applies to monkeys who have learned to recognize the symbols of Arabic numerals.[33]

When we subtract two numbers, say, 9 − 6, the time that we take is directly proportional to the size of the subtracted number[34]—so it takes longer to perform 9 − 6 than, say, 9 − 4 or 9 − 2. Everything happens as if we have to mentally move along the number line, starting from the first number and taking as many steps as the second number: the further we have to go, the longer we take. We do not crunch symbols like a digital computer; instead, we use a slow and serial spatial metaphor, motion along the number line. Likewise, when we think of a price, we cannot help but attribute to it a fuzzier value when the number gets larger—a remnant of our primate-based number sense, whose precision decreases with number size.[35] This is why, against all rationality, when we negotiate, we are ready to give up a few thousand dollars on the price of an apartment and, the same day, bargain a few quarters on the price of bread: the level of imprecision that we tolerate is proportional to a number's value, for us just as for macaques.

And the list goes on: parity, negative numbers, fractions . . . all these concepts are demonstrably grounded in the representation of quantities that we inherit from evolution.[36] Unlike a digital computer, we are unable to manipulate symbols in the abstract: we always grind them in concrete and often approximate quantities. The persistence of such analog effects in an educated brain betrays the ancient roots of our concept of numbers.

Approximate numbers are one of the old pillars on which the construction of mathematics is founded. However, education also leads to considerable enrichment of this original number concept. When we learn to count and calculate, the mathematical symbols that we acquire allow us to perform precise computations. This is a revolution: for millions of years, evolution had been content with fuzzy quantities. Symbol learning is a powerful factor for change: with education, all our brain circuits are repurposed to allow for the manipulation of exact numbers.

Number sense is certainly not the only foundation of mathematics. As we saw earlier, we also inherit from our evolution a sense of space, with its own specialized neural circuits containing place, grid, and head direction cells. We also have a sense of shape, which allows any toddler to distinguish rectangles, squares, and triangles. In a way that is not yet fully understood, under the influence of symbols such as words and numbers, all these concepts are recycled when we learn mathematics. The human brain manages to recombine them, in a language of thought, in order to form new concepts.[37] The basic building blocks that we inherit from our evolutionary history become the foundational primitives of a new, productive language in which mathematicians write new pages every day.

READING RECYCLES THE CIRCUITS OF VISION AND SPOKEN LANGUAGE

What about learning to read? Reading is yet another example of neuronal recycling: to read, we reuse a vast set of brain areas that are initially dedicated to vision and spoken language. In my book *Reading in the Brain*,[38] I describe, in detail, the circuits of literacy. When we learn to read, a subset of our visual regions becomes specialized in recognizing strings of letters and sends them to spoken language areas. As a result, in any good reader, written words end up being processed exactly as spoken words: literacy creates a new visual gateway to our language circuits.

Long before children learn how to read, they obviously possess a sophisticated visual system that allows them to recognize and name objects, animals, and people. They can recognize any image regardless of its size, position, or orientation in 3-D space, and they know how to associate a name to it. Reading recycles part of this preexisting picture naming circuit. The acquisition of literacy involves the emergence of a region of the visual cortex

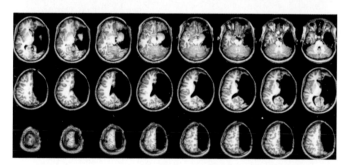

Figure 1.
Brain plasticity can sometimes overcome major obstacles. At the age of three, Nico's right hemisphere was surgically removed (see MRI slices in the middle). Yet this major loss did not prevent him from becoming an accomplished artist, capable of painting both excellent copies (bottom) and original work (top). Learning squeezed all his talents, including language, math, reading, and painting, into a single hemisphere.

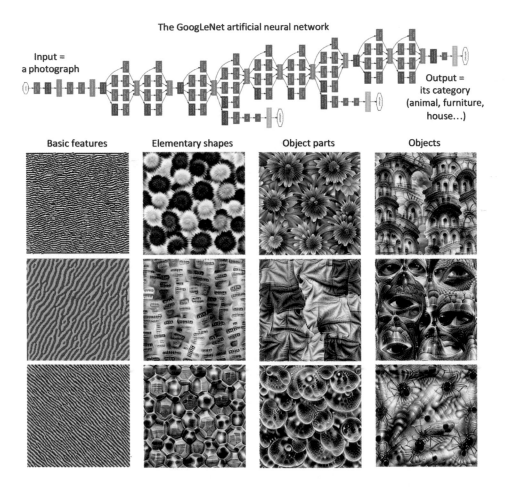

The GoogLeNet artificial neural network

Input =
a photograph

Output =
its category
(animal, furniture,
house...)

| Basic features | Elementary shapes | Object parts | Objects |

Figure 2.

Learning means developing a hierarchy of representations appropriate to the problem at hand. In the GoogLeNet network, which learns to identify images, the adjustment of millions of parameters allows each level of the hierarchy to recognize a useful aspect of reality. At the lowest level, the simulated neurons are sensitive to basic features such as oriented lines or textures. As we climb the hierarchy, neurons respond to increasingly complex shapes, including houses, eyes, and insects.

Figure 3.

How does a deep neural network learn to categorize handwritten numbers? This is a difficult problem because a given digit can be written in hundreds of different ways. At the lowest level of the neuronal hierarchy (bottom right), the artificial neurons confuse numbers that look alike, such as 9 and 4. The higher up in the hierarchy we go, the more successful neurons are in grouping all images of the same number and separating them by clear boundaries.

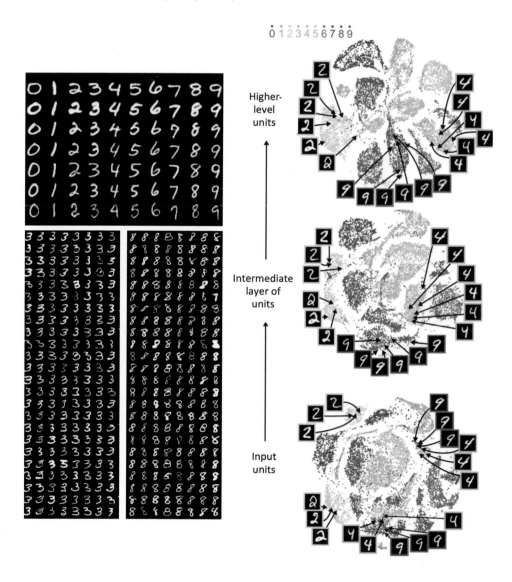

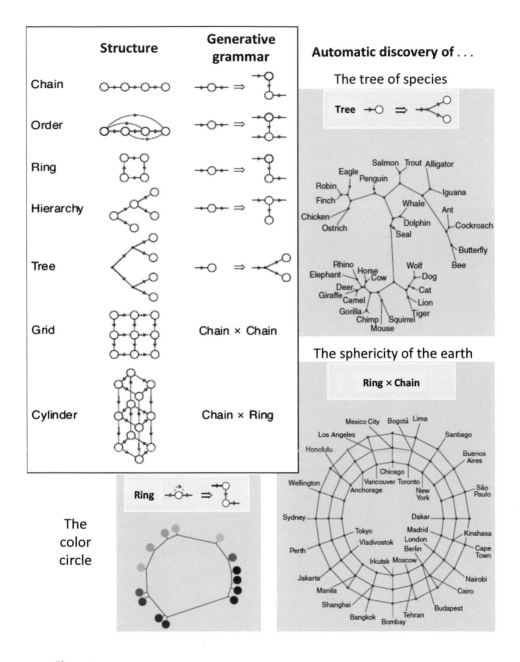

Figure 4.

Learning means inferring the grammar of a domain. At MIT, two computer scientists invented an algorithm that discovers the hidden structure of a scientific field. The system is endowed with a grammar of rules whose combinations generate all kinds of new structures: lines, planes, circles, cylinders.... By selecting the structure that best fits the data, the algorithm makes discoveries that took scientists years: the tree of animal species (Darwin, 1859), the roundness of the earth (Parmenides, 600 BCE), and the circle of colors (Newton, 1675).

Figure 5.
Far from being blank slates, babies possess vast amounts of knowledge. In the laboratory, researchers uncover the sophistication of babies' intuitions by measuring their surprise when they are subjected to situations that violate the laws of physics, arithmetic, probability, or geometry.

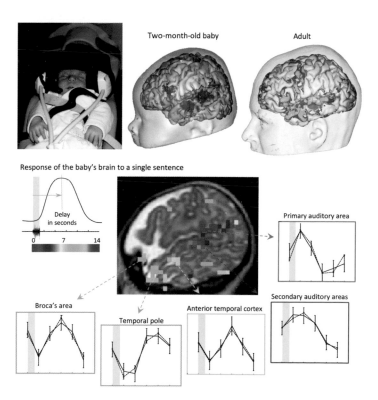

Figure 6.
At birth, the infant brain already channels spoken language into specific circuits of the left hemisphere. When babies are scanned using functional MRI while listening to sentences in their mother tongue, a specific network of brain regions lights up—the same as in adults. The activity starts in the primary auditory area, then gradually extends to the temporal and frontal areas, in the same order as in the adult brain. These data refute the idea of an initially disorganized brain, a mere blank slate that awaits the imprint of its environment.

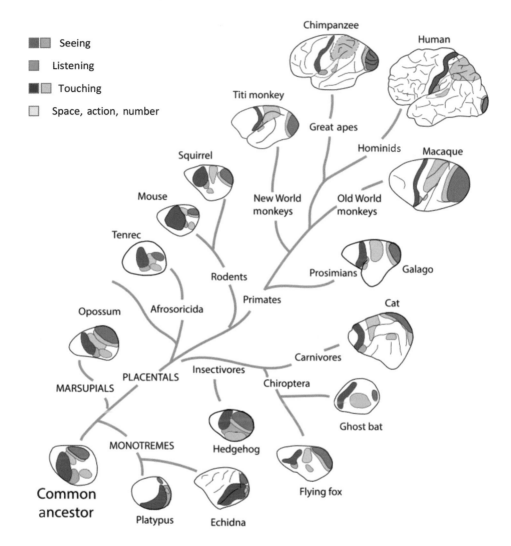

Seeing

Listening

Touching

Space, action, number

Figure 7.

The architecture of the human brain has a long evolutionary history. Many specialized regions (here, the primary sensory areas) share their basic layout with other species. They are wired in utero, under the influence of many genes, and are already active during the third trimester of pregnancy. The primate brain is characterized by proportionally smaller sensory areas and an enormous expansion of the cognitive regions of the parietal (gray), temporal, and especially prefrontal cortex. In *Homo sapiens*, these regions are remarkably plastic: they shelter a language of thought and enable us to increase our knowledge throughout life.

Figure 8.

In the first weeks of pregnancy, the body organizes itself on a genetic basis. No learning is necessary for the five fingers to form and receive their specific innervation. Similarly, the fundamental architecture of the brain is laid down in the absence of any learning. At birth, the cortex is already organized, folded, and connected in a manner which is common to all human beings, and which distinguishes us from all other primates. The detailed wiring, however, is free to vary depending on the environment. By the third trimester of gestation, the fetal brain already begins to adapt to the information it receives from the outside world.

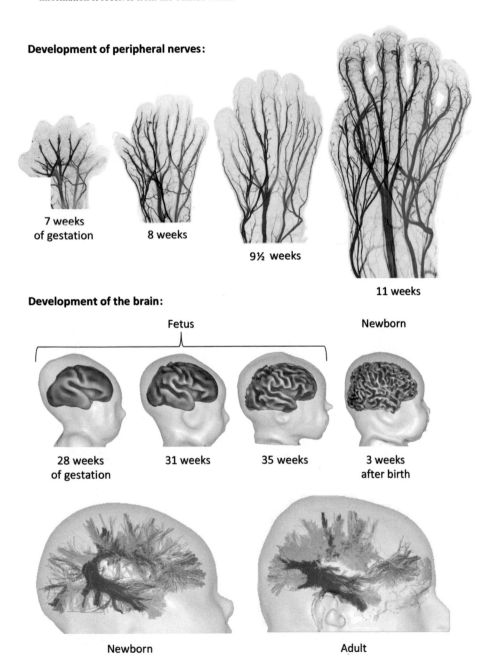

Development of peripheral nerves:

7 weeks
of gestation

8 weeks

9½ weeks

11 weeks

Development of the brain:

Fetus

Newborn

28 weeks
of gestation

31 weeks

35 weeks

3 weeks
after birth

Newborn

Adult

Figure 9.

The human cortex is subdivided into specialized areas. As early as 1909, the German neurologist Korbinian Brodmann (1868–1918) noted that the size and distribution of neurons vary across the different regions of the cortex. For instance, within Broca's area, which is involved in language processing, Brodmann delineated three areas (numbered 44, 45, and 47). These distinctions have been confirmed and refined by molecular imaging. The cortex is tiled with distinct areas whose boundaries are marked by sudden variations in neurotransmitter receptor density. During pregnancy, certain genes are selectively expressed in the different regions of the cortex and help subdivide it into specialized organs.

Brodmann
map of cortical areas (1909)

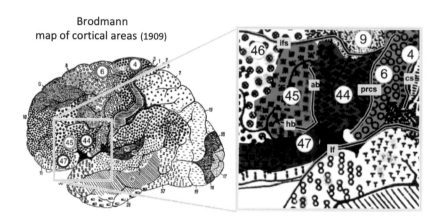

Borders between cortical areas, as defined by four receptor molecules

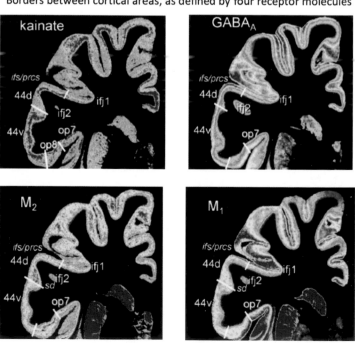

Neurons from the entorhinal cortex:

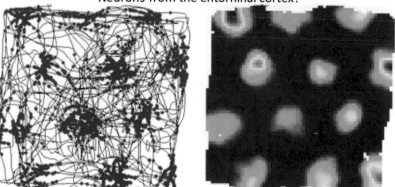

Figure 10.
When a physical system self-organizes, be it lava or beeswax, it is not uncommon for hexagons to form. The nervous system is no exception: in a region of the entorhinal cortex, which acts as the GPS of the brain, neurons self-organize into "grid cells" that tile the physical space with a lattice of triangles and hexagons. When a rat explores a large room, each neuron fires only when the animal lies at the vertex of one of those triangles. Such grid cells appear a single day after the mouse begins to move around: the sense of space is based on an almost innate GPS circuit.

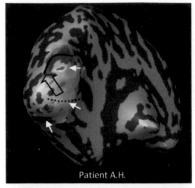

Patient A.H.

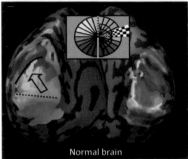

Normal brain

Figure 11.
Synaptic plasticity allows the brain to partially reorganize itself when it suffers serious damage. The patient A.H. (top) was born with only one cerebral hemisphere: at seven weeks of gestation, her right hemisphere stopped developing. In a normal brain (bottom), the early visual areas of the left hemisphere represent only the right half of the world (colored blue and green in the central disc). However, in patient A.H., very small regions reorganized and began to respond to the left half of the world (in red, indicated by white arrows). Thus, A.H. is not totally blind on the left side, unlike an adult who suffered the same lesion. Nevertheless, this reorganization is modest: in the primary visual cortex, genetic determinism trumps brain plasticity.

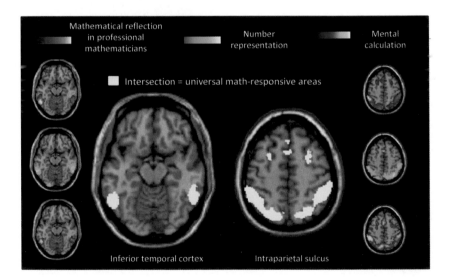

Mathematical reflection in professional mathematicians

Number representation

Mental calculation

Intersection = universal math-responsive areas

Inferior temporal cortex

Intraparietal sulcus

Figure 12.
Education consists of recycling ancient brain circuits, thus redirecting them toward new functions. Since infancy, we all possess areas for representing numbers (in green), which we also use for mental calculation (in blue). Remarkably, even professional mathematicians continue to use the same brain regions when thinking about higher-level math concepts (in red). These neural networks initially respond to concrete sets of objects, but later get recycled for more abstract concepts.

Figure 13.

The acquisition of mathematics is largely independent of sensory experience. Even the blind can become superb mathematicians—and in them, the same regions of parietal, temporal, and frontal cortex are activated during mathematical reflection as in sighted mathematicians. The only difference is that they also recycle their visual cortex to do math.

NICHOLAS SAUNDERSON.
Lucasian Professor of Mathematicks

Fifteen sighted mathematicians:

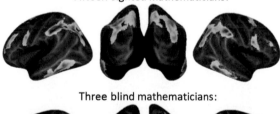

Three blind mathematicians:

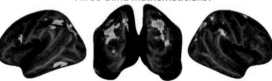

Additional activation of visual cortex in the blind:

Experiment 1   Experiment 2

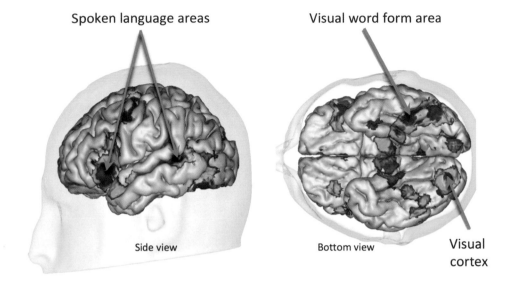

Spoken language areas   Visual word form area

Side view   Bottom view   Visual cortex

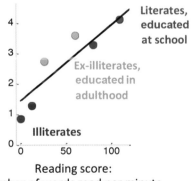

Brain response to written sentences

Literates, educated at school

Ex-illiterates, educated in adulthood

Illiterates

Reading score: number of words read per minute

Figure 14.

Learning to read recycles a network of brain areas involved in vision and spoken language. The regions in color are those affected by reading acquisition: their activity in response to a written sentence increases with reading score, from pure illiterates to expert readers. Literacy affects the brain in two different ways: it specializes the visual areas for written letters, particularly in a region of the left hemisphere called the "visual word form area," and it activates the circuits of spoken language through vision.

Figure 15.

Functional MRI can be used to track the acquisition of literacy in children. As soon as a child learns to read, a visual region of the left hemisphere starts to specialize for letter strings. Reading recycles part of the mosaic of regions that all primates use to recognize faces, objects, and places.

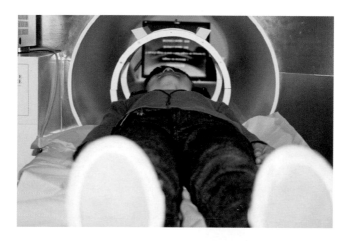

Six-year-old non-readers

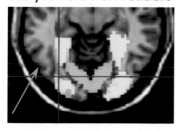

Six-year-old readers

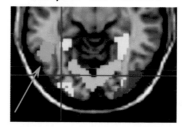

Nine-year-old dyslexics

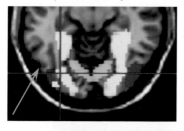

Nine-year-old readers

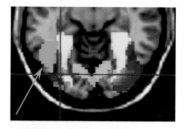

Response to Words

Figure 16.
Alerting signals can massively modulate learning. Neuromodulators such as serotonin, acetylcholine, and dopamine, whose signals are broadcast to much of the cortex, tell us when to pay attention and seem to force the brain to learn. In the experiment shown at the bottom, rats listened to a nine-kilohertz sound which was associated with an electrical stimulation of the basal nucleus of Meynert, thus triggering the release of acetylcholine in the cortex. After a few days of exposure, the entire auditory cortex was invaded by this sound frequency and its neighbors (regions in blue).

Cerebral circuits of acetylcholine

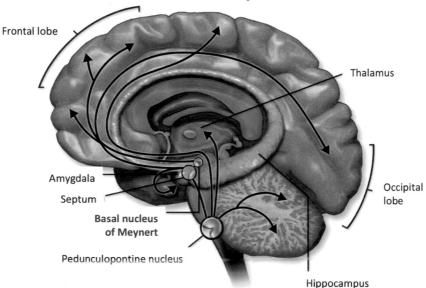

Frontal lobe

Thalamus

Amygdala

Septum

Basal nucleus of Meynert

Pedunculopontine nucleus

Occipital lobe

Hippocampus

Modulation of learning

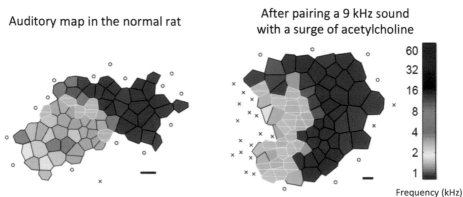

Auditory map in the normal rat

After pairing a 9 kHz sound with a surge of acetylcholine

60
32
16
8
4
2
1

Frequency (kHz)

Auditory cortex: detection of a deviant sound

Frequent sounds

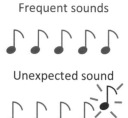

Unexpected sound

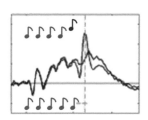

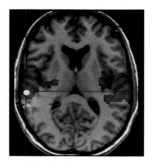

Prefrontal cortex: detection of a violation in the melody

Frequent melody

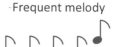

Unexpected melody

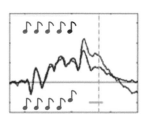

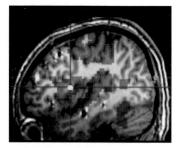

Figure 17.
Error feedback is the third pillar of learning. By detecting and correcting its errors, the brain progressively learns to adjust its models of the environment. Virtually all brain regions emit and exchange error signals. In this experiment, the brain learns to detect violations in a sequence of sounds. First, a short melody of five notes is played several times. When the sequence changes without warning, a surprise response (in red) signals the error to other regions of the brain and allows them to amend their predictions. Auditory areas react to local violations of expectations (top), while an extensive network, which includes the prefrontal cortex, responds to global violations of the entire melody (bottom).

Figure 18.

Consolidation is the fourth pillar of learning. Initially, all learning requires considerable effort, accompanied by intense activation of the parietal and frontal regions for spatial and executive attention. For a beginner reader, for instance, deciphering words is a slow, effortful, and sequential process: the more letters a word has, the slower the child reads (top). With practice, automaticity arises: reading becomes a fast, parallel, and unconscious process (bottom). A specialized reading circuit emerges, freeing up cortical resources for other tasks.

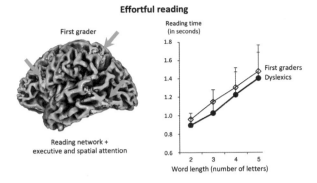

Effortful reading

First grader

Reading network + executive and spatial attention

Reading time (in seconds)

First graders
Dyslexics

Word length (number of letters)

Automatic reading

Same child toward the end of second grade

Specialized reading network

Reading time (in seconds)

Second graders
Third graders

Word length (number of letters)

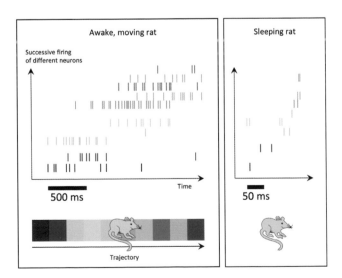

Figure 19.

Sleep plays an important role in the consolidation of learning. When a rat falls asleep, the neurons in its hippocampus replay, often at an accelerated speed, the very same sequences of activity that it experienced when it was awake. This activity, which extends to the cortex, may be repeated hundreds of times during the night. Such neuronal replay helps consolidate and automatize what was learned during the previous day. While we sleep, our brain may even discover regularities that eluded us the day before.

that my colleague Laurent Cohen and I have dubbed the "visual word form area." This region concentrates our learned knowledge of letter strings, to such an extent that it can be considered as our brain's "letter box." It is this brain area, for instance, that allows us to recognize a word regardless of its size, position, font, or cAsE, whether UPPERCASE or lowercase.[39] In any literate person, this region, which is located in the same spot in all of us (give or take a few millimeters), serves a dual role: it first identifies a string of learned characters, and then, through its direct connections to language areas,[40] it allows those characters to be quickly translated into sound and meaning.

What would happen if we scanned an illiterate child or adult as she progressively learned to read? If the theory is correct, then we should literally see her visual cortex reorganize. The neuronal recycling theory predicts that reading should invade an area of the cortex normally devoted to a similar function and repurpose it to this novel task. In the case of reading, we expect a competition with the preexisting function of the visual cortex, which is to recognize all sorts of objects, bodies, faces, plants, and places. Could it be that we lose some of the visual functions that we inherited from our evolution as we learn to read? Or, at the very least, are these functions massively reorganized?

This counterintuitive prediction is precisely what my colleagues and I tested in a series of experiments. To draw a complete map of the brain regions that are changed by literacy, we scanned illiterate adults in Portugal and Brazil, and we compared them to people from the same villages who had had the good fortune of learning to read in school, either as children or adults.[41] Unsurprisingly perhaps, the results revealed that, with reading acquisition, an extensive map of areas had become responsive to written words (see figure 14 in the color insert). Flash a sentence, word by word, to an illiterate individual, and you will find that their brain does not respond much: activity spreads to early visual areas, but it stops there, because the letters cannot be recognized. Present the same sequence of written words to an adult who has learned to read, and a much more extended cortical circuit now lights up, in direct proportion to the person's reading score. The areas activated include the letter box area, in the left occipitotemporal cortex, as well as all the classical language regions associated with language comprehension. Even the earliest visual areas increase their response: with reading acquisition, they seem to become attuned to the recognition of small print.[42] The more fluent a person is, the more these regions are activated by written words, and the more they strengthen their links: as reading becomes increasingly automatic, the translation of letters into sounds speeds up.

But we can also ask the opposite question: Are there regions that are more active among bad readers and whose activity decreases as one learns to read? The answer is positive: in illiterates, the brain's responses to faces are more intense. The better we read, the more this activity decreases in the left hemisphere, at the exact place in the cortex where written words find their niche—the brain's letter box area. It's as if the brain needs to make room for letters in the cortex, so the acquisition of reading interferes with the prior function of this region, which is the recognition of faces and objects. But, of course, since we do not forget how to recognize faces when we learn to read, this function is not just chased out of the cortex. Rather, we have also observed that, with literacy, the response to faces increases in the right hemisphere. Driven out of the left hemisphere, which is the seat of language and reading for most of us, faces take refuge on the other side.[43]

We first made this observation in literate and illiterate adults, but we quickly replicated our results in children who were learning to read.[44] As soon as a child begins to read, the visual word form area begins to respond in the left hemisphere. Meanwhile its symmetrical counterpart, in the right hemisphere, strengthens its response to faces (see figure 15 in the color insert). The effect is so powerful that, for a given age, just by examining the brain activity evoked by faces, a computer algorithm can correctly decide whether a child has or has not yet learned to read. And when a child suffers from dyslexia, these regions do not develop normally—neither on the left, where the visual word form area fails to emerge, nor on the right, where the fusiform cortex fails to develop a strong response to faces.[45] Reduced activity of the left

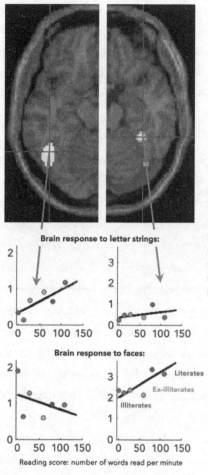

Brain response to letter strings:

Brain response to faces:

Literates
Ex-illiterates
Illiterates

Reading score: number of words read per minute

In agreement with the neuronal recycling hypothesis, learning to read competes with the previous functions of the visual cortex—in this case, face recognition. With increasing levels of literacy, from pure illiterates to expert readers, the activation evoked by written words increases in the left hemisphere—and the activation evoked by faces moves from the left hemisphere to the right.

occipitotemporal cortex to written words is a universal marker of reading difficulties in all countries where it has been tested.[46]

Recently, we got permission to conduct a bold experiment. We wanted to see the reading circuits emerge in individual children—and to this aim, we brought the same children back to our brain-imaging center every two months, from the end of kindergarten through the end of first grade. The results lived up to our expectations. The first time we scanned these children, there was not much to be seen: as long as the children had not yet learned how to read, their cortex responded selectively to objects, faces, and houses, but not to letters. After two months of schooling, however, a specific response to written words appeared, at the same exact location as in adults: the left occipito-temporal cortex. Very slowly, the representation of faces changed: as the children became more and more literate, face responses increased in the right hemisphere, in direct proportion to reading scores. Once again, in agreement with the neuronal recycling hypothesis, we could see reading acquisition compete with the prior function of the left occipitotemporal cortex, the visual recognition of faces.

We realized while doing this work that this competition could be explained in two different ways. The first possibility is what we called the "knockout model": from birth on, faces settle in the visual cortex of the left hemisphere, and learning to read later knocks them straight out into the right hemisphere. The second possibility is what we termed the "blocking model": the cortex develops slowly, gradually growing specialized patches for faces, places, and objects; and when letters enter this developing landscape, they take over part of the available territory and prevent the expansion of other visual categories.

So, does literacy lead to a knockout or a blockade of the cortex? Our experiments suggest the latter: learning to read blocks the growth of face-recognition areas in the left hemisphere. We witnessed this blockade thanks to the MRI scans that we acquired every two months from the children who were learning to read.[47] At this age, around six or seven, cortical specialization is still far from complete. A few patches are already dedicated to faces, objects, and places, but there are also many cortical sites that have not yet specialized for any given category. And we could visualize their progressive specialization: when children entered first grade and quickly began to read, letters invaded one of those poorly specified regions and recycled it. Contrary to what I initially thought, letters do not completely overrun a preexisting face patch; they move in right next door, in a free sector of cortex, a bit like an aggressive supermarket setting up shop right next to a small grocery store. The expansion of one stops the other—and because letters settle down in the left hemisphere,

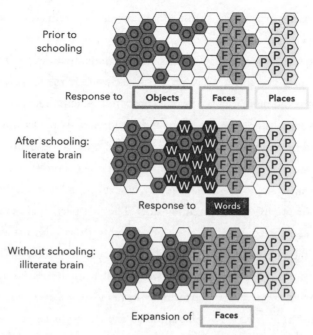

Learning is easier in childhood, while the cortex is still malleable. Before a young child goes to school, some visual regions of the brain have already specialized in recognizing objects, faces, and places—but there are still large patches with little or no specialization (symbolized by empty hexagons). Learning to read invades these labile circuits and blocks the growth of other categories of objects. If a child does not learn to read, those regions become involved in recognizing faces and objects, and gradually lose their ability to learn letters.

which is dominant for language, faces have no choice but to move to the right side.

In brief, the ventral visual system is still undergoing major reorganization during the early school years. The fact that our schools typically teach children to read between the ages of six and eight nicely dovetails with the evidence for intense brain plasticity during this time period. We have organized our education system so that it efficiently takes advantage of a sensitive period when the visual cortex is particularly pliable. While its overall architecture is highly constrained from birth, the human inferotemporal cortex possesses the remarkable ability to adapt to various shapes and learn all kinds of images. When exposed to thousands of written words, this region recycles itself for this new activity, in a specific sector which happens to be innately connected to language circuits.

As we get older, our visual cortex seems to gradually freeze and lose the ability to tune to new images. The progressive closure of the sensitive period makes it more and more difficult for the cortex to efficiently recognize letters and their combinations. My colleagues and I studied two people who tried to learn to read as adults: one of them had never had the chance to go to school, while the other had suffered a small stroke in the visual word form area, rendering him fully "alexic"—unable to read. We scanned them regularly for two years.[48] Their progress was incredibly slow. The first participant eventually developed a specialized region for letters, but this growth did not affect the face area—the circuits for face recognition had been imprinted in his brain and seemed no longer able to move. Our stroke patient, on the other hand, never managed to re-create a new "letter box" in his visual cortex. His reading improved but remained slow and similar to the laborious deciphering of a novice reader—being an adult, he was missing the neuronal plasticity necessary to recycle part of his cortex into an automated reading machine.

MUSIC, MATH, AND FACES

The conclusion is simple: to profoundly recycle our visual cortex and become excellent readers, we must take advantage of the period of maximum plasticity that early childhood offers. Our research shows several other examples. Take musical reading: a musician who learned to read sheet music at an early age has practically double the surface area of his visual cortex dedicated to musical symbols, compared with someone who has never learned music.[49] This massive growth occupies space on the surface of the cortex, and it seems to dislodge the visual word form area from its usual place: in musicians, the cortical region that responds to letters, the brain's letter box, is displaced by nearly one centimeter from its normal position in nonmusicians.

Another example is our varying abilities to decode mathematical equations.

An accomplished mathematician must be able to recognize, at a glance, expressions as obscure as $\pi = 3.141592...$, $\varphi = 1.61803394...$, $f(x) = a_0 + \sum_{n=1}^{\infty} \left(a_n \cos \frac{n\pi x}{L} + b_n \sin \frac{n\pi x}{L} \right)$, or $e^x = 1 + \frac{x}{1!} + \frac{x^2}{2!} + \frac{x^3}{3!} + \cdots$, just like we read a sentence in a novel. I once attended a conference where the brilliant French mathematician Alain Connes (another Fields Medal winner) exhibited an extraordinarily dense equation that was twenty-five lines long. He explained that this all-encompassing mathematical expression captured all the physical effects of all known elementary particles. A second mathematician pointed his finger and said, "Isn't there an error on line thirteen?" "No," Connes immediately answered without losing his composure, "because the corresponding compensating term is right there on line fourteen!"

How is this remarkable knack for complex formulas reflected in the brains of mathematicians? Brain imaging shows that these mathematical objects invade the lateral occipital regions of both hemispheres—after math training, these regions respond to algebraic expressions much more so than in non-mathematicians. And, once again, we witness a competition with faces: this time, the patches of face-responsive cortex wane away in both hemispheres.[50] In other words, while literacy merely drives faces out of the left hemisphere and forces them to move over to the right hemisphere, intense practice with numbers and equations interferes with the representation of faces on both sides, leading to a global shrinkage of the visual face-recognition circuitry.

It is tempting to relate this finding to the famous myth of the eccentric mathematician, uninterested in anything other than his equations and unable to recognize his neighbor, his dog, or even his reflection in the mirror. There is, indeed, an abundance of anecdotes and jokes about mindless mathematicians. For instance, what's the difference between an introvert mathematician and an extrovert mathematician? While he's talking to you, the introvert looks at his shoes. But the extrovert mathematician looks at *your* shoes! . . .

In reality, we do not yet know whether the reduction in cortical responses to faces in math buffs is directly related to their supposed lack of social competence (which, I should say, is more of a myth than a reality—many mathematicians are wonderfully at ease in society). Most crucially, causality remains to be determined: Does spending one's life in mathematical formulas reduce the response to faces? Or, on the contrary, do mathematicians immerse themselves in a universe of equations because they find them easier than social interactions? Whatever the answer may be, cortical competition is a genuine phenomenon, and the representation of faces in our brains turns out to be remarkably sensitive to education and schooling, to the point where it can provide a reliable marker of whether a child has received training in math, music, or reading. Neuronal recycling is a reality.

THE BENEFITS OF AN ENRICHED ENVIRONMENT

The take-home message is that both sides of the nature-nurture debate are right: a child's brain is both structured and plastic. At birth, all children are equipped with a panoply of specialized circuits, shaped by genes, themselves selected by tens of millions of years of evolution. This self-organization gives the baby's brain a deep intuition of several major areas of knowledge: a sense of the physics governing objects and their motion; a knack for spatial navigation; intuitions of numbers, probability, and mathematics; an inclination toward other human beings; and even a genius for languages—the blank-slate metaphor could not be more wrong. And yet evolution also left the door open to many learning opportunities. Not everything is predetermined in the child's brain. Quite the contrary: the detail of neural circuits, on a scale of a few millimeters, is largely open to interactions with the outside world.

During the first years of life, genes guide an exuberant overproduction of neural circuits: twice as many synapses as necessary. In a way that we do not fully understand yet, this initial abundance opens up an immense space of mental models of the world. The brains of young children swarm with possibilities and explore a much wider set of hypotheses than the brains of adults. Each baby is open to all languages, all scripts, all possible mathematics—within the genetic limits of our species, of course.

And the baby's brain also comes equipped with another innate gift: powerful learning algorithms that select the most useful synapses and circuits, thus providing a second layer of adaptation of the organism to its environment. Thanks to them, as early as the first few days of life, the brain begins to specialize and settle into its configuration. The first regions to freeze are the sensory areas: early visual areas mature in a few years, and it takes less than twelve months before the auditory areas converge toward the vowels and consonants of the child's native language. As the sensitive periods of brain plasticity close, one after the other, a few years suffice for any of us to become a native of a given language, writing, and culture. And if we are deprived of stimulation in a certain domain, whether we are orphans in Bucharest or illiterates in the suburbs of Brasilia, we risk forever losing our mental flexibility in this field of knowledge.

This is not to say that intervention is not worth the effort, at any age: the brain retains some of its plasticity throughout its life, especially in its highest-level regions such as the prefrontal cortex. However, everything points to the optimal effectiveness of early intervention. Whether the goal is to make an owl wear glasses, teach an adopted child a second language, or help a child adjust to deafness, blindness, or the loss of a whole cerebral hemisphere, the sooner, the better.

Our schools are institutions designed to make the most of the plasticity of the developing brain. Education relies heavily on the spectacular flexibility of the child's brain to recycle some of its circuits and reorient them toward new activities such as reading or mathematics. When schooling begins early, it can transform lives: numerous experiments show that children from disadvantaged backgrounds who benefit from early educational interventions show improved outcomes, even decades later, in many domains—from lower crime rates to higher IQs and incomes to better health.[51]

But schooling is not a magic pill. Parents and families also have a duty to stimulate children's brains and enrich their environments as much as possible. All babies are budding physicists who love to experiment with gravity and falling bodies—as long as they are allowed to tinker, build, fail, and start over again, rather than being strapped in a car seat for hours. All children are nascent mathematicians who love counting, measuring, drawing lines and circles, assembling shapes—provided one gives them rulers, compasses, paper, and attractive math puzzles. All infants are genial linguists: as early as eighteen months of age, they easily acquire ten to twenty words a day—but only if they are spoken to. Their families and friends must feed this appetite for knowledge and nourish them with well-formed sentences, without hesitating to use a rich lexicon. Many studies show that a child's vocabulary at three to four years old directly depends on the amount of child-directed speech they received during their first years. Passive exposure does not suffice: active one-to-one interactions are essential.[52]

All research findings are remarkably convergent: enriching the environment of a young child helps her build a better brain. For instance, in children who are read bedtime stories every evening, the brain circuits for spoken language are stronger than in other toddlers—and the strengthened cortical pathways are precisely those that will later allow them to understand texts and formulate complex thoughts.[53] Likewise, children who are lucky enough to be born into bilingual families, with each parent giving them the wonderful gift of speaking in their native language, effortlessly acquire two lexicons, two grammars, and two cultures—at no cost.[54] Throughout their lives, their bilingual brains retain better abilities for language processing and for acquiring a third or fourth language. And when they enter old age, their brains seem to resist the ravages of Alzheimer's disease for longer. Exposing the developing brain to a stimulating environment allows it to keep more synapses, larger dendrites, and more flexible and redundant circuits[55]—like the owl that learned to wear prism glasses and kept, for its entire life, more diversified dendrites and a greater ability to switch from one behavior to another. Let's diversify our children's early learning portfolio: the blossoming of their brains depends in part on the richness of the stimulation they receive from their environment.

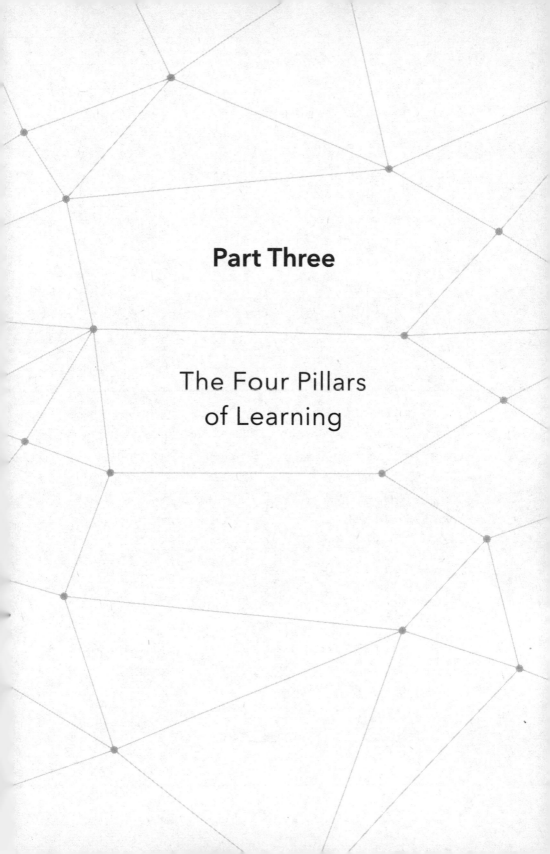

Part Three

The Four Pillars
of Learning

The mere existence of synaptic plasticity does not suffice to explain the extraordinary success of our species. Indeed, such plasticity is omnipresent in the animal world: even house flies, nematode worms, and sea slugs have modifiable synapses. If *Homo sapiens* became *Homo docens*, if learning became our ecological niche and the main reason behind our global success, it is because the human brain contains a whole bag of additional tricks.

During evolution, four major functions appeared that maximized the speed with which we extracted information from our environment. I call them the four pillars of learning, because each of them plays an essential role in the stability of our mental constructions: if even one of these pillars is missing or weak, the whole structure quakes and quivers. Conversely, whenever we need to learn, and learn fast, we can rely on them to optimize our efforts. These pillars are:

- Attention, which amplifies the information we focus on.
- Active engagement, an algorithm also called "curiosity," which encourages our brain to ceaselessly test new hypotheses.
- Error feedback, which compares our predictions with reality and corrects our models of the world.

- Consolidation, which renders what we have learned fully automated and involves sleep as a key component.

Far from being unique to humans, these functions are shared with many other species. However, due to our social brain and language skills, we exploit them more effectively than any other animal—especially in our families, schools, and universities.

Attention, active engagement, error feedback, and consolidation are the secret ingredients of successful learning. And these fundamental components of our brain architecture are deployed both at home and at school. Teachers who manage to mobilize all four functions in their students will undoubtedly maximize the speed and efficiency with which their classes can learn. Each of us should therefore learn to master them.

CHAPTER 7

Attention

IMAGINE ARRIVING AT THE AIRPORT JUST IN TIME TO CATCH A PLANE. Everything in your behavior betrays the heightened concentration of your attention. Your mind on alert, you look for the departures sign, without letting yourself be distracted by the flow of travelers; you quickly scroll through the list to find your flight. Advertisements all around call out to you, but you do not even see them—instead, you head straight for the check-in counter. Suddenly, you turn around: in the crowd, an unexpected friend just called your first name. This message, which your brain considers a priority, takes over your attention and invades your consciousness . . . making you forget which check-in counter you were supposed to go to.

In the space of a few minutes, your brain went through most of the key states of attention: vigilance and alertness, selection and distraction, orientation and filtering. In cognitive science, "attention" refers to all the mechanisms by which the brain selects information, amplifies it, channels it, and deepens its processing. These are ancient mechanisms in evolution: whenever a dog reorients its ears or a mouse freezes up upon hearing a cracking sound, they're making use of attention circuits that are very close to ours.[1]

Why did attention mechanisms evolve in so many animal species? Because attention solves a very common problem: information saturation. Our brain is constantly bombarded with stimuli: the senses of sight, hearing, smell, and touch transmit millions of bits of information per second. Initially, all these messages are processed in parallel by distinct neurons—yet it would be impossible to digest them in depth: the brain's resources would not suffice. This is why a pyramid of attention mechanisms, organized like a gigantic filter, carries out a selective triage. At each stage, our brain decides how much importance it should attribute to such and such input and allocates resources only to the information it considers most essential.

Selecting relevant information is fundamental to learning. In the absence of attention, discovering a pattern in a pile of data is like looking for the fabled needle in a haystack. This is one of the main reasons behind the slowness of conventional artificial neural networks: they waste considerable time analyzing all possible combinations of the data provided to them, instead of sorting out the information and focusing on the relevant bits. It was only in 2014 that two researchers, Canadian Yoshua Bengio and Korean Kyunghyun Cho, showed how to integrate attention into artificial neural networks.[2] Their first model learned to translate sentences from one language to another. They showed that attention brought in immense benefits: their system learned better and faster because it managed to focus on the relevant words of the original sentence at each step.

Very quickly, the idea of learning to pay attention spread like wildfire in the field of artificial intelligence. Today, if artificial systems manage to successfully label a picture ("A woman throwing a Frisbee in a park"), it is because they use attention to channel the information by focusing a spotlight on each relevant part of the image. When describing the Frisbee, the network concentrates all its resources on the corresponding pixels of the image and temporarily removes all those which correspond to the person and the park—it will return to them later.[3] Nowadays, any sophisticated artificial intelligence system no longer connects all inputs with all outputs—it knows that learning will be faster if such a plain network, where every pixel of the input has a chance to predict any word at the output, is replaced by an organized architecture where learning is broken down into two modules: one that learns to pay attention, and another that learns to name the data filtered by the first.

A woman throwing a <u>Frisbee</u> in a park.

A little <u>girl</u> sitting on a bed with
a teddy bear.

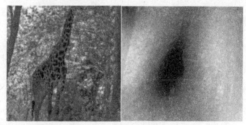

A giraffe standing in a forest with
<u>trees</u> in the background.

The first pillar of learning is attention, a mechanism so fundamental that it is now being integrated into most contemporary artificial neural networks. Here, the machine learns to find the words to describe an image. Selective attention acts as a spotlight that lights up certain areas of the image (in white on the right) and discards everything else. At any given moment, attention thus concentrates all the learning power on a selected data set.

Attention is essential, but it may result in a problem: if attention is misdirected, learning can get stuck.[4] If I don't pay attention to the Frisbee, this part of the image is wiped out: processing goes on as if it did not exist. Information about it is discarded early on, and it remains confined to the earliest sensory areas. Unattended objects cause only a modest activation that induces little or no learning.[5] This is utterly different from the extraordinary amplification that occurs in our brain whenever we pay attention to an object and become aware of it. With conscious attention, the discharges of the sensory and conceptual neurons that code for an object are massively amplified and prolonged, and their messages propagate into the prefrontal cortex, where whole populations of neurons ignite and fire for a long time, well beyond the original duration of the image.[6] Such a strong surge of neural firing is exactly what synapses need in order to change their strength—what neuroscientists call "long-term potentiation." When a pupil pays conscious attention to, say, a foreign-language word that the teacher has just introduced, she allows that word to deeply propagate into her cortical circuits, all the way into the prefrontal cortex. As a result, that word has a much better chance of being remembered. Unconscious or unattended words remain largely confined to the brain's sensory circuits, never getting a chance to reach the deeper lexical and conceptual representations that support comprehension and semantic memory.

This is why every student should learn to pay attention—and also why teachers should pay more attention to attention! If students don't attend to the right information, it is quite unlikely that they will learn anything. A teacher's greatest talent consists of constantly channeling and capturing children's attention in order to properly guide them.

Attention plays such a fundamental role in the selection of relevant information that it is present in many different circuits in the brain. American psychologist Michael Posner distinguishes at least three major attention systems:

1. Alerting, which indicates <u>when</u> to attend, and adapts our level of vigilance.
2. Orienting, which signals <u>what</u> to attend to, and amplifies any object of interest.
3. Executive attention, which decides <u>how</u> to process the attended information, selects the processes that are relevant to a given task, and controls their execution.

These systems massively modulate brain activity and can therefore facilitate learning, but also point it in the wrong direction. Let us examine them one by one.

ALERTING: THE AWAKENING OF THE BRAIN

The first attention system, perhaps the oldest in evolution, tells us *when* to be on the watch. It sends warning signals that mobilize the entire body when circumstances require it. When a predator approaches or when a strong emotion overwhelms us, a whole series of subcortical nuclei immediately increases the wakefulness and vigilance of the cortex. This system dictates a massive and diffuse release of neuromodulators such as serotonin, acetylcholine, and dopamine (see figure 16 in the color insert). Through long-range axons with many spread-out branches, these alerting messages reach virtually the entire cortex, greatly modulating cortical activity and learning. Some researchers speak of a "now print" signal, as if these messages directly tell the cortex to commit the current contents of neural activity into memory.

Animal experiments show that the firing of this warning system can indeed radically alter cortical maps (see figure 16 in the color insert). The American neurophysiologist Michael Merzenich conducted several experiments in which the alerting system of mice was tricked into action by electrical stimulation of their subcortical dopamine or acetylcholine circuits. The outcome was a massive shift in cortical maps. All the neurons that happened to be activated at that moment, even if they had no objective importance, were subject to intense amplification. When a sound, for instance, a high-pitched tone, was systematically associated with a flash of dopamine or acetylcholine, the mouse's brain became heavily biased toward this stimulus. As a result, the whole auditory map was invaded by this arbitrary note. The mouse became better and better at discriminating sounds close to this sensitive note, but it partially lost the ability to represent other frequencies.[7]

It is remarkable that such cortical plasticity, induced by tampering with the alerting system, can occur even in adult animals. Analysis of the circuits involved shows that neuromodulators such as serotonin and acetylcholine—particularly via the nicotinic receptor (sensitive to nicotine, another major player in arousal and alertness)—modulate the firing of cortical inhibitory interneurons, tipping the balance between excitation and inhibition.[8] Remember that inhibition plays a key role in the closing of sensitive periods for synaptic plasticity. Disinhibited by the alerting signals, cortical circuits seem to recover some of their juvenile plasticity, thus reopening the sensitive period for signals that the mouse brain labels as crucial.

What about *Homo sapiens*? It is tempting to think that a similar reorganization of cortical maps occurs every time a composer or a mathematician passionately dives into their chosen field, especially when their passion starts at an early age. A Mozart or a Ramanujan is perhaps so electrified by fervor that his brain maps become literally invaded with mental models of music or math. Furthermore, this may apply not only to geniuses, but to anyone passionate in their work, from a manual worker to a rocket scientist. By allowing cortical maps to massively reshape themselves, passion breeds talent.

Even though not everyone is a Mozart, the same brain circuits of alertness and motivation are present in all people. What circumstances of daily life would mobilize these circuits? Do they activate only in response to trauma or strong emotions? Maybe not. Some research suggests that video games, especially action games that play with life and death, provide a particularly effective means of engaging our attentional mechanisms. By mobilizing our alerting and reward systems, video games massively modulate learning. The dopamine circuit, for example, fires when we play an action game.[9] Psychologist Daphné Bavelier has shown that this translates into rapid learning.[10] The most violent action games seem to have the most intense effects, perhaps because they most strongly mobilize the brain's alerting circuits. Ten hours of gameplay suffice to improve visual detection, refine the rapid estimation of the number of objects on the screen, and expand the capacity to concentrate on a target without being distracted. A video game player manages to make ultrafast decisions without compromising his or her performance.

Parents and teachers complain that today's children, plugged into computers, tablets, consoles, and other devices, constantly zap from one activity to the next and have lost the capacity to concentrate—but this is untrue. Far from reducing our ability to concentrate, video games can actually increase it. In the future, will they help us remobilize synaptic plasticity in adults and children alike? Undoubtedly, they are a powerful stimulant of attention, which is why my laboratory has developed a whole range of educational tablet games for math and reading, based on cognitive science principles.[11]

Video games also have their dark side: they present well-known risks of social isolation, time loss, and addiction. Fortunately, there are many other ways to unlock the effects of the alerting system while also drawing on the brain's social sense. Teachers who captivate their students, books that draw in their readers, and films and plays that transport their audiences and immerse them in real-life experiences probably provide equally powerful alerting signals that stimulate our brain plasticity.

ORIENTING: THE BRAIN'S FILTER

The second attention system in the brain determines *what* we should attend to. This orienting system acts as a spotlight on the outside world. From the millions of stimuli that bombard us, it selects those to which we should allocate our mental resources, because they are urgent, dangerous, appealing . . . or merely relevant to our present goals.

The founding father of American psychology, William James (1842–1910), in his *The Principles of Psychology* (1890), best defined this function of attention: "Millions of items of the outward order are present to my senses which never properly enter into my experience. Why? Because they have no interest for me. My experience is what I agree to attend to. Only those items which I notice shape my mind."

Selective attention operates in all sensory domains, even the most abstract. For example, we can pay attention to the sounds around us: dogs move their ears, but for us humans, only an internal pointer in our brain moves and tunes in to whatever we decide to focus on. At a noisy cocktail party, we are able to select one out of ten conversations based on voice and meaning. In vision, the orienting of attention is often more obvious: we generally move our head and eyes toward whatever attracts us. By shifting our gaze, we bring the object of interest into our fovea, which is an area of very high sensitivity in the center of our retina. However, experiments show that even without moving our eyes, we can still pay attention to any place or any object, wherever it is, and amplify its features.[12] We can even attend to one of several superimposed drawings, just like we attend to one of several simultaneous conversations. And there is nothing stopping you from paying attention to the color of a painting, the shape of a curve, the speed of a runner, the style of a writer, or the technique of a painter. Any representation in our brains can become the focus of attention.

In all these cases, the effect is the same: the orienting of attention amplifies whatever lies in its spotlight. The neurons that encode the attended information increase their firing, while the noisy chattering of other neurons is squashed. The impact is twofold: attention makes the attended neurons more sensitive to the information that we consider relevant, but, above all, it increases their influence on the rest of the brain. Downstream neural circuits echo the stimulus to which we lend our eyes, ears, or mind. Ultimately, vast expanses of cortex reorient to encode whatever information lies at the center of our attention.[13] Attention acts as an amplifier and a selective filter.

"The art of paying attention, the great art," says the philosopher Alain (1868–1951), "supposes the art of not paying attention, which is the royal art." Indeed, paying attention also involves choosing what to ignore. For an object to come into the spotlight, thousands of others must remain in the shadows. To direct attention is to choose, filter, and select: this is why cognitive scientists speak of selective attention. This form of attention amplifies the signal which is selected, but it also dramatically reduces those that are deemed irrelevant. The technical term for this mechanism is "biased competition": at any given moment, many sensory inputs compete for our brain's resources, and attention biases this competition by strengthening the representation of the selected item while squashing the others. This is where the spotlight metaphor reaches its limits: to better light up a region of the cortex, the attentional spot-light of our brain also reduces the illumination of other regions. The mechanism relies on interfering waves of electrical activity: to suppress a brain area, the brain swamps it with slow waves in the alpha frequency band (between eight and twelve hertz), which inhibit a circuit by preventing it from developing coherent neural activity.

Paying attention, therefore, consists of suppressing the unwanted information—and in doing so, our brain runs the risk of becoming blind to what it chooses not to see. Blind, really? Really. The term is fully appropriate, because many experiments, including the famous "invisible gorilla" experiment,[14] demonstrate that inattention can induce a complete loss of sight. In this classic experiment, you are asked to watch a short movie where basketball players, dressed in black and white, pass a ball back and forth. Your task is to count, as precisely as you can, the number of passes of the white team. A piece of cake, you think—and indeed, thirty seconds later, you triumphantly give the right answer. But now the experimenter asks a strange question: "Did you see the gorilla?" The gorilla? What gorilla? We rewind the tape, and to your amazement, you discover that an actor in a full-body gorilla costume walked across the stage and even stopped in the middle to pound on his chest for several seconds. It seems impossible to miss. Furthermore, experiments show that, at some point, your eyes looked right at the gorilla. Yet you did not see it. The reason is simple: your attention was entirely focused on the white team and therefore actively inhibited the distracting players who were dressed in black . . . gorilla included! Busy with the counting task, your mental workspace was unable to become aware of this incongruous creature.

The invisible gorilla experiment is a landmark study in cognitive science, and one which is easily replicated: in a great variety of settings, the mere act of focusing our attention blinds us to unattended stimuli. If, for instance, I ask you to judge whether the pitch of a sound is high or low, you may become blind to another stimulus, such as a written word that appears within the next fraction of a second. Psychologists call this phenomenon the "attentional blink":[15] your eyes may remain open, but your mind "blinks"—for a short while, it is fully busy with its main task and utterly unable to attend to anything else, even something as simple as a single word.

In such experiments, we actually suffer from two distinct illusions. First, we fail to see the word or the gorilla, which is bad enough. (Other experiments show that inattention can lead us to miss a red light or run over a pedestrian—never use your cell phone behind the wheel!) But the second illusion is even worse: we are unaware of our own unawareness—and, therefore, we are absolutely convinced that we have seen all there is to see! Most people who try the invisible gorilla experiment cannot believe their own blindness. They think that we played a trick on them, for instance by using two different movies. Typically, their reasoning is that if there really was a gorilla in the video, they would have seen it. Unfortunately, this is false: our attention is extremely limited, and despite all our good will, when our thoughts are focused on one object, other objects—however salient, amusing, or important—can completely elude us and remain invisible to our eyes. The intrinsic limits of our awareness lead us to overestimate what we and others can perceive.

The gorilla experiment truly deserves to be known by everyone, especially parents and teachers. When we teach, we tend to forget what it means to be ignorant. We all think that what we see, everyone can see. As a result, we often have a hard time understanding why a child, despite the best of intentions, fails to *see*, in the most literal sense of the term, what we are trying to teach him. But the gorilla heeds a clear message: seeing requires attending. If students, for one reason or another, are distracted and fail to pay attention, they may be entirely oblivious to their teacher's message—and what they cannot perceive, they cannot learn.[16]

As an example, consider an experiment recently performed by the American psychologist Bruce McCandliss which probed the role of attention in learning to read.[17] Is it better to pay attention to the individual letters of a word or to the overall form of the whole word? To find out, McCandliss and his colleagues taught adults an unusual writing system made up of elegant curves. The subjects were first trained with sixteen words, then their brain responses were recorded while they tried to read these sixteen learned words, as well as sixteen new words in the same script. Unbeknownst to them, however, their attention was also being manipulated. Half the participants were told to attend to the curves as a whole, because each of them, much like a Chinese character, corresponded to one word. The other group was told that, in fact, the curves were made up of three superimposed letters, and that they would learn better by paying attention to each letter. Thus, the first group paid attention on the whole-word level, while the second group attended to the individual letters, which had actually been used to write the words.

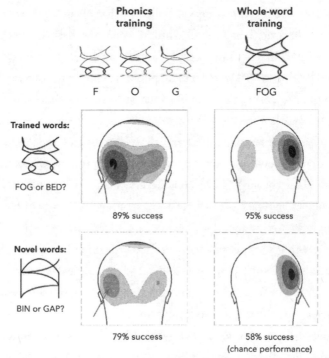

Selective attention can orient learning to the right or wrong circuit. In this experiment, adults learned to read a new writing system using either a phonics approach or a whole-word approach. Those who attended to the overall shape of the words did not realize that the words were made of letters, even after three hundred trials. Whole-word attention directed the learning to an inappropriate circuit in the right hemisphere and prevented the participants from generalizing what they had learned to novel words. When attention was drawn to the presence of letters, however, people were able to decipher the alphabet and to read novel words, using the normal reading circuit located in the left ventral visual cortex.

What were the results? Both groups managed to remember the first sixteen words, but attention radically altered their ability to decipher new words.
The participants in the second group, focused on letters, discovered many of
the correspondences between letters and sounds and were able to read 79
percent of the new words. Furthermore, an examination of their brains showed
that they had activated the normal reading circuitry, localized to the ventral
visual areas of the left hemisphere. In the first group, however, attending to
the overall word form completely hindered the capacity to generalize to novel
items: these volunteers could not read any new words, and they activated a
totally inappropriate circuit located in the visual areas of the *right* hemisphere.

The message is clear: attention radically changes brain activity. Paying
attention to the overall shape of the words prevents the discovery of the alpha-
betic code and directs brain activity toward an inadequate circuit in the
opposite hemisphere. To learn to read, phonics training is essential. Only by
attending to the correspondence between letters and sounds can a student
activate the classical reading circuit, allowing for the proper type of learning
to take place. All first-grade teachers who teach reading should be familiar
with this data: they show how important it is to properly direct children's
attention. Many converging data convincingly demonstrate the superiority of
such a phonics approach over whole-word reading.[18] When a child attends to
the letter level, for instance, by tracking each letter with her finger, from left
to right, learning becomes much easier. If, on the other hand, the child is not
provided with any attentional clues and naively examines the written word as
a whole, without attending to its internal structure, nothing happens. Attention is a key ingredient of successful learning.

Above all, therefore, good teaching requires permanent attention to children's attention. Teachers must carefully choose what they want children to
attend to, because only the items that lie at the focus of attention are represented in the brain with sufficient strength to be efficiently learned. The other
stimuli, the losers of the attentional competition, cause little or no stir inside
the child's plastic synapses.

The efficient teacher therefore pays close attention to his pupils' mental states. By constantly stirring children's curiosity with attention-grabbing lessons, he ensures that each class is a memorable experience. By tailoring his teaching to each child's attention span, he ensures that all students follow the entire lesson.

EXECUTIVE CONTROL: THE BRAIN'S SWITCHBOARD

Our third and final attention system determines *how* the attended information is processed. The executive control system, sometimes called the "central executive," is a hodgepodge of circuits that allows us to choose a course of action and stick to it.[19] It involves a whole hierarchy of cortical areas, mainly located in the frontal cortex—the huge mass of cortex that lies beneath our forehead and comprises close to a third of the human brain. Compared with other primates, our frontal lobes are enlarged, better connected, and packed with a larger number of neurons, each with a broader and more complex dendritic tree.[20] It's no wonder, then, that human cognitive abilities are much more developed than those of any other primate—and this is especially true at the highest level of the cognitive hierarchy, which allows us to supervise our mental operations and become aware of our mistakes: the executive control system.[21]

Imagine having to mentally multiply 23 by 8. It is your executive control system that ensures that the whole series of relevant mental operations runs smoothly from beginning to end: first, focus on the ones digit (3) and multiply it by 8, then store the result (24) in memory; now focus on the tens digit (2) and also multiply it by 8 to obtain 16, and remember that you are working in the tens column, therefore it corresponds to 160; and finally, add 24 and 160 to reach the final result: 184.

Executive control is the switchboard of the brain: it orients, directs, and governs our mental processes, much like a railroad yardman who tends the switches in a busy railway station and manages to bring each train to the right track by choosing the appropriate orientation for each switch. The brain's central executive is considered one of the attention systems because, like the others, it selects from many possibilities—but this time, from the available mental operations rather than from the stimuli that reach us. Thus, spatial attention and executive attention complement each other. When we do mental arithmetic, spatial attention is the system that scans the mathematics textbook page and shines the spotlight on the problem 23 × 8—but it is executive attention which then guides the spotlight step by step, first selecting the 3 and the 8, then routing them to the brain circuits for multiplication, and so on. The central executive activates the relevant operations and inhibits the inappropriate ones. It constantly ensures that the mental program runs smoothly, and decides when to change strategies. It is also the system which, within a specialized subcircuit of the cingulate cortex, detects when we make an error, or when we deviate from the goal, and immediately corrects our action plan.

There is a close link between executive control and what cognitive scientists call working memory. In order to follow a mental algorithm and control its execution, we must constantly keep in mind all the elements of the ongoing program: intermediate results, steps already carried out, operations remaining to be performed. . . . Thus, executive attention controls the inputs and the outputs of what I have called the "global neural workspace": a temporary conscious memory within which we can maintain, for a short period, practically any piece of information that seems relevant to us and relay it to any other module.[22] The global workspace acts as the brain's router, the signalman that decides how, and in what order, to send the information to the many different processors that our brain hosts. At this level, mental operations are slow and serial: this is a system that processes one piece of information at a time and is therefore incapable of doing two operations at once. Psychologists also call it the "central bottleneck."

Are we really unable to execute two mental programs at once? We are sometimes under the impression that we can simultaneously perform two tasks, or even follow two distinct trains of thought—but this is a pure illusion. A basic experiment illustrates this point: Give someone two very simple tasks—for example, pressing a key with the left hand whenever they hear a high-pitched sound, and pressing another key with the right hand if they see the letter Y. When both targets occur simultaneously or in close succession, the person performs the first task at a normal speed, but the execution of the second task is considerably slowed down, in direct proportion to the time spent making the first decision.[23] In other words, the first task delays the second: while our global workspace is busy with the first decision, the second one has to wait. And the lag is huge: it easily reaches a few hundred milliseconds. If you are too concentrated on the first task, you may even miss the second task entirely. Remarkably, however, none of us is aware of this large dual-task delay—because, by definition, we cannot be aware of information before it enters our conscious workspace. While the first stimulus gets consciously processed, the second one has to wait outside the door, until the global workspace is free—but we have no introspection of that waiting time, and if asked about it, we think that the second stimulus appeared exactly when we were finished with the first, and that we processed it at a normal speed.[24]

Once again, we are unaware of our mental limits (indeed, it would be paradoxical if we could somehow become aware of our lack of awareness!). The only reason we believe that we can multitask is that we are unaware of the huge delay it causes. Thus, many of us continue to text while we drive—in spite of all the evidence that texting is one of the most distracting activities ever. The lure of the screen and the myth of multitasking are among the most dangerous fabrications of our digital society.

What about training? Can we ever turn ourselves into genuine multitaskers who do multiple things at once? Perhaps, but only with intense training on one of the two tasks. Automatization frees the conscious workspace: by routinizing an activity, we can execute it unconsciously, without tying up the brain's central resources. Through hard practice, for instance, a professional pianist may be able to talk while playing, or a typist may be able to copy a document while listening to the radio. However, these are rare exceptions, and psychologists continue to debate them, because it is also possible that executive attention quickly switches from one task to the next in an almost undetectable manner.[25] The basic rule stands: in any multitask situation, whenever we have to perform multiple cognitive operations under the control of attention, at least one of the operations is slowed down or forgotten altogether.

Because of this severe effect of distraction, learning to concentrate is an essential ingredient of learning. We cannot expect a child or an adult to learn two things at once. Teaching requires paying attention to the limits of attention and, therefore, carefully prioritizing specific tasks. Any distraction slows down or wastes our efforts: if we try to do several things at once, our central executive quickly loses track. In this respect, cognitive science experiments in the lab converge nicely with educational findings. For instance, field experiments demonstrate that an overly decorated classroom distracts children and prevents them from concentrating.[26] Another recent study shows that when students are allowed to use their smartphones in class, their performance suffers, even months later, when they are tested on the specific content of that day's class.[27] For optimal learning, the brain must avoid any distraction.

LEARNING TO ATTEND

Executive attention roughly corresponds to what we call "concentration" or "self-control." Importantly, this system is not immediately available to children: it will take fifteen or twenty years before their prefrontal cortex reaches its full maturity. Executive control emerges slowly throughout childhood and adolescence as our brain, through experience and education, gradually learns to control itself. Much time is needed for the brain's central executive to systematically select the appropriate strategies and inhibit the inadequate ones, all the while avoiding distraction.

Cognitive psychology is full of examples where children gradually correct their mistakes as they increasingly manage to concentrate and inhibit inappropriate strategies. Psychologist Jean Piaget was the first to notice this: Very young children sometimes make seemingly silly mistakes. If, for example, you hide a toy a few times at location A, and then switch to hiding it at location B, babies below one year of age continue to look for it at location A (even if they saw perfectly well what happened). This is the famous "A-not-B error," which led Piaget to conclude that infants lack object permanence—the knowledge that an object continues to exist when it is hidden. However, we now know that this interpretation is wrong. Examination of the babies' eyes shows that they know where the hidden object is. But they have trouble resolving mental conflicts: in the A-not-B task, the routine response that they learned on previous trials tells them to go to location A, while their more recent working memory tells them that, on the present trial, they should inhibit this habitual response and go to location B. Before ten months of age, the habit prevails. At this age, what is lacking is executive control, not object knowledge. Indeed, the A-not-B error disappears around twelve months of age, in direct relation to the development of the prefrontal cortex.[28]

Another typical error of children is the confusion between number and size. Here again, Piaget made an essential discovery but got the interpretation wrong. He found that young children, before they were about three years old, had trouble judging the number of objects in a group. In his classical number conservation experiments, Piaget first showed children two equal rows of marbles, in one-to-one correspondence, such that even the youngest children would agree that the rows had the same number of marbles. He would then space the marbles in one of the rows apart:

oooooo   oooooo   →   o o o o o o   oooooo

Remarkably, the children would now affirm that the two sets were unequal, and that the longer row had more objects. This is a surprisingly silly error—but contrary to what Piaget thought, it does not mean that children at this age are incapable of "conserving number." As we have seen, even newborn babies already possess an abstract sense of number, independent of the spacing of items or even the sensory modality in which they are presented. No, the difficulty arises, once again, from executive control. Children must learn to inhibit a prominent feature (size) and amplify a more abstract one (number).

Even in adults, such selective attention may fail. For instance, we all have a hard time deciding which of two sets is larger when the items in the smaller set are bigger and more spread out in space; and we also have a hard time choosing the larger number between 7 and 9. What develops with age and education is not so much the intrinsic precision of the number system, but the ability to use it efficiently without getting distracted by irrelevant cues, such as density or size.[29] Once again, progress in such tasks correlates with the development of neural responses in the prefrontal cortex.[30]

I could multiply the examples: at all stages of life and in all domains of knowledge, whether cognitive or emotional, it is primarily the development of our executive control abilities which allows us to avoid making errors.[31] Let's try it on your own brain: name the *color of the ink* (black or white) in which each of the following words is printed:

> dog house well because sofa too
> white black white black white black

When you reached the second half of the list, did the task become more difficult? Did you slow down and make errors? This classic effect (which is even more striking when the words are printed in color) reflects the intervention of your executive control system. When the words and colors conflict, the central executive must inhibit word reading to remain focused on the task of naming the ink color.

Now try solving the following problem: "Mary has twenty-six marbles. This is four more than Gregory. How many marbles does Gregory have?" Did you have to fight the urge to add the two numbers? Did you think of thirty instead of the correct result of twenty-two? The problem statement uses the word "more" even though you have to subtract—this is a trap that many children fall into before they manage to control themselves and think deeper about the meanings of such math problems in order to select the relevant arithmetic operation.

Attention and executive control develop spontaneously with the progressive maturation of the prefrontal cortex, which extends over the first two decades of our lives. But this circuit, like all others, is plastic, and many studies show that its development can be enhanced by training and education.[32] Because this system intervenes in a great variety of cognitive tasks, many educational activities, including the most playful, can effectively develop executive control. The American psychologist Michael Posner was the first to develop educational software that improves young children's ability to concentrate. One game, for instance, forces the player to heed the orientation of a fish in the center of the screen. The target fish is surrounded by others that face in the opposite direction. In the course of the game, which consists of many levels of increasing difficulty, the child progressively learns to avoid being distracted by the target fish's neighbors—a simple task that teaches concentration and inhibition. This is just one of many ways to encourage reflection and discourage immediate, knee-jerk responding.

Long before computers were invented, the Italian doctor and teacher Maria Montessori (1870–1952) noticed how a variety of practical activities could develop concentration in young children. In today's Montessori schools, for example, children walk along an ellipse drawn on the ground, without ever taking their feet off the line. Once they succeed, the difficulty is raised by having them walk with a spoon in their mouth, then with a ping-pong ball in the spoon, and so on. Experimental studies suggest that the Montessori approach has a positive impact on many aspects of child development.[33] Other studies demonstrate the attentional benefits of video games, meditation, or the practice of a musical instrument. . . . For a young child, controlling their body, gaze, and breathing while coordinating their two hands can be an excruciatingly difficult task—that is probably why playing music at an early age has a strong impact on the attention circuits of the brain, including a significant bilateral increase in the thickness of the prefrontal cortex.[34]

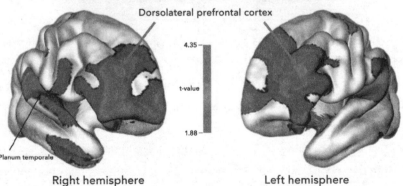

Executive attention, the ability to concentrate and control oneself, develops with age and education. Learning to play a musical instrument is one of the many ways to enhance concentration and self-control from an early age. The cortex is thicker in musicians than in well-matched nonmusicians, particularly the dorsolateral prefrontal cortex, which plays an important role in executive control.

Training in executive control can even change one's IQ. This may come as a surprise, because IQ is often viewed as a given—a fundamental determinant of children's mental potential. However, the intellectual quotient is just a behavioral ability, and as such, it is far from being unchangeable by education. Like any of our abilities, IQ rests on specific brain circuits whose synaptic weights can be changed by training. What we call fluid intelligence—the ability to reason and solve new problems—makes massive use of the brain's executive control system: both mobilize a similar network of brain areas, notably the dorsolateral prefrontal cortex.[35] Indeed, standardized measures of fluid intelligence resemble the tests that cognitive psychologists use to assess executive control: both emphasize attention, concentration, and the ability to move quickly from one activity to another, without losing sight of the overall goal. And in fact, training programs that focus on working memory and executive control cause a slight increase in fluid intelligence.[36] These results are consistent with previous findings showing that although intelligence is not devoid of genetic determinism, it can change dramatically in response to environmental factors, including education. And these effects can be enormous. In one study, low-IQ children between the ages of four and six were adopted in families with either a high or a low socioeconomic status. At adolescence, those who had landed in the better-off families had gained twenty IQ points, compared to only eight points for the others.[37] A recent meta-analysis examined the effect of education on intelligence, and concluded that each additional year at school yields a gain of one to five IQ points.[38]

The current frontier of research involves optimizing the effects of cognitive training and clarifying their limits. Can the effects last for years? How can we ensure that they extend well beyond the trained tasks, in various situations throughout life? This is the challenge, because, by default, the brain tends to develop tricks specific to each task, on a case-by-case basis. The solution probably lies in the diversification of learning experiences, and the best results seem to be obtained by educational programs that stimulate the core cognitive skills of working memory and executive attention in a great variety of contexts.

Certain findings make me particularly optimistic. Early training in working memory, especially if done in kindergarten, appears to have positive effects on concentration and success in many areas, including those most directly relevant to school: reading and mathematics.[39] This is not surprising, since we have known for years that working memory is one of the best predictors of later success in arithmetic.[40] The effects of these exercises are multiplied if we combine memory training with more direct teaching of the concept of the "number line"—the essential idea that numbers are organized on a linear axis where adding or subtracting consists of moving right or left.[41] All these educational interventions seem to be the most beneficial to children from disadvantaged backgrounds. For families at low socioeconomic levels, early intervention, starting in kindergarten and teaching the fundamentals of learning and attention, can be one of the best educational investments.

I'LL ATTEND IF YOU ATTEND

ὁ ἄνθρωπος φύσει πολιτικὸν ζῷον
Man is by nature a social (or political) animal.

Aristotle (350 BCE)

All mammalian species—including, of course, all primates—possess attention systems. But attention in humans exhibits a unique feature that further accelerates learning: social attention sharing. In *Homo sapiens*, more than in any other primate, attention and learning depend on social signals: I attend where you attend, and I learn from what you teach me.

From the earliest age, infants gaze at faces and pay particular attention to people's eyes. As soon as something is said to them, their first reflex is not to explore the scene, but to catch the gaze of the person they are interacting with. Only once eye contact is established do they turn toward the object that the adult is staring at. This remarkable ability for social attention sharing, also called "shared attention," determines what children learn.

I have already told you about experiments where babies are taught the meaning of a new word, such as "wog." If the infants can follow the speaker's gaze toward the so-called wog, they have no trouble learning this word in just a few trials—but if wog is repeatedly emitted by a loudspeaker, in direct relation to the same object, no learning occurs. The same goes for learning phonetic categories: a nine-month-old American child who interacts with a Chinese nanny for only a few weeks acquires Chinese phonemes—but if he receives exactly the same amount of linguistic stimulation from a very high-quality video, no learning occurs.[42]

Hungarian psychologists Gergely Csibra and György Gergely postulate that teaching others and learning from others are fundamental evolutionary adaptations of the human species.[43] *Homo sapiens* is a social animal whose brain is endowed with circuits for "natural pedagogy" that are triggered as soon as we attend to what others are trying to teach us. Our global success is due, at least in part, to a specific evolutionary trait: the ability to share attention with others. Most of the information we learn, we owe to others, rather than to our personal experience. In this manner, the collective culture of the human species can rise far above what any individual could discover alone. This is what psychologist Michael Tomasello calls the "cultural ratchet" effect: like a ratchet prevents an elevator from falling back down, social sharing prevents culture from regressing. Whenever one person makes a useful discovery, it quickly spreads to the whole group. Thanks to social learning, it is very rare for the cultural elevator to come down and for a major invention to be forgotten.

Our attentional system has adapted to this cultural context. Gergely and Csibra's research shows that, from an early age, children's attention is highly attuned to adult signals. The presence of a human tutor, who looks at the child before making a specific demonstration, massively modulates learning. Not only does eye contact attract the child's attention, but it also signals that the tutor intends to teach the child an important point. Even babies are sensitive to this: eye contact puts them in a "pedagogical stance" that encourages them to interpret the information as important and generalizable.

Let's take an example: A young woman turns to object A with a big smile,

then to object B with a grimace. An eighteen-month-old baby watches the
scene. What conclusion will the baby draw? It all depends on the signals that
the child and the adult exchanged. If no eye contact was established, then the

Shared attention:

| An adult looks at the child . . . | . . . then turns toward an object while smiling . . . | . . . and turns toward another object while frowning. | Another person makes a request. | A majority of children give her the preferred object. |

Comprehension of intentions:

An adult, her hands **occupied**, makes a strange action with her head.

An adult, her hands **free**, makes a strange action with her head.

80% of children imitate the action intelligently with their **hands.**

70% of children imitate the action faithfully with their **hands.**

Social interactions are an essential ingredient of the human learning algorithm. What we learn
depends on our understanding of the intentions of others. Even eighteen-month-old babies understand that if you look them in the eye, you are trying to convey important information to them.
Following eye contact, they learn more effectively and succeed more in generalizing than other
people (top). As early as fourteen months of age, babies can already interpret people's intentions:
after seeing a person turn on a light with her head, they imitate this gesture in every way, unless
the person's hands were occupied, in which case babies understand that they can simply press the
button with their hands (bottom).

child simply remembers one specific piece of information: this person likes object A and dislikes object B. If, however, eye contact was established, then the child deduces much more: he believes that the adult was trying to teach him something important, and he therefore draws the more general conclusion that object A is pleasant and object B is bad, not only for this person in particular but for everyone. Children pay extreme attention to any evidence of voluntary communication. When someone gives obvious signs of trying to communicate with them, they infer that this person wants to teach them abstract information, not just their own idiosyncratic preferences.

It is not only eye contact that matters: children also quickly understand the communicative intention that lies behind the act of pointing with a finger (whereas chimpanzees never really understand this gesture). Even babies realize when someone is trying to get their attention and give them important information. For instance, when nine-month-old babies see someone trying to catch their attention and then pointing to an object, they later remember the identity of that object, because they understand that this is the information that matters to their interlocutor—whereas, if they see the same person reaching toward the object without looking at them, they remember only the position of the object, not its identity.[44]

Parents and teachers, always keep this crucial fact in mind: your attitude and your gaze mean everything for a child. Getting a child's attention through visual and verbal contact ensures that she shares your attention and increases the chance that she will retain the information you are trying to convey.

TEACHING IS ATTENDING TO SOMEONE ELSE'S KNOWLEDGE

No other species can teach like we do. The reason is simple: we are probably the only animals with a theory of other people's minds, an ability to pay attention to them and imagine their thoughts—including what they think others think, and so on and so forth, in an infinite loop. This type of recursive representation is typical of the human brain and plays an essential role in the pedagogical relationship. Educators must constantly think about what their pupils do not know: teachers adapt their words and choose their examples in order to increase their students' knowledge as quickly as possible. And the pupils know that their teacher knows that they do not know. Once children adopt this pedagogical stance, they interpret each act of the teacher as an attempt to convey knowledge to them. And the loop goes on forever: adults know that children know that adults know that they do not know . . . which allows adults to choose their examples knowing that children will try to generalize them.

This pedagogical relationship may well be unique to *Homo sapiens*: it does not seem to exist in any other species. In 2006, a landmark article[45] published in *Science* described a form of teaching in the meerkat, a small South African mammal of the mongoose family—but in my view, the study misused the very definition of teaching. What was it about? The biggest family affair: learning how to prepare food! Mongooses face a serious cooking challenge: they feed on extremely dangerous prey, scorpions with deadly stingers that need to be removed before eating. Their plight is similar to that of Japanese cooks preparing fugu, a fish whose liver, ovaries, eyes, and skin contain deadly doses of the paralyzing drug tetrodotoxin: one error in the recipe, and you are dead. Japanese chefs train for three years before they are allowed to serve their first fugu—but how do meerkats acquire their know-how? The *Science* paper convincingly showed that adult meerkats help their young by first offering them "prepared" food consisting of scorpions with the stingers removed. As young meerkats grow, the adults provide them with an increasing proportion of live scorpions, and this obviously helps the young become independent hunters. Thus, according to the authors, three teaching criteria are met: the adult performs a specific behavior in the presence of the young; this behavior has a cost for the adult; and the young benefit by acquiring knowledge more quickly than if the adult had not intervened.

The case of meerkats is certainly noteworthy: during mongoose evolution, a singular mechanism emerged that clearly facilitates survival. But is this genuine teaching? In my opinion, the data do not allow us to conclude that meerkats really teach their young, because one crucial ingredient is missing:

shared attention to one another's knowledge. There is no evidence that adult meerkats pay any attention to what the young know or, conversely, that the young take into account the pedagogical stance of the adults. Adult mongooses only present increasingly dangerous prey to their young as they age. As far as we know, this mechanism could be completely pre-wired and specific to scorpion consumption—a complex but narrow-minded behavior comparable to the famous bee dance or the flamingo's bridal parade.

In brief, although we attempt to project onto mongooses and scorpions our own preconceptions, a closer look reveals how far their behavior is from ours. With its obvious limitations, the story of the teaching mongoose actually teaches us, as in a negative image, what is truly unique and precious about our species. The genuine pedagogical relationships that happen in our schools and universities involve strong mental bonds between teachers and students. A good teacher builds a mental model of his students, their skills and their mistakes, and takes every action to enrich his pupils' minds. This ideal definition therefore excludes any teacher (human or computer) who merely mechanically delivers a stereotypical lesson, without tailoring it to the prior knowledge and expectations of his audience—such mindless, unidirectional teaching is inefficient. On the flip side, teaching is efficient only when the students, for their part, have good reasons to be persuaded that teachers do their best to convey their knowledge. Any healthy pedagogical relationship must be based on bidirectional streams of attention, listening, respect, and mutual trust. There is currently no evidence that such a "theory of mind"—the capacity of students and teachers to attend to one another's mental states—exists in any animal other than the human species.

The meerkat's modest pedagogy also fails to do justice to the role that education plays in human societies. "Every man is a humanity, a universal history," says Jules Michelet (1798–1874). Through education, we convey to others the best thoughts of the thousands of human generations that preceded us. Every word, every concept we learn is a small conquest that our ancestors passed on to us. Without language, without cultural transmission, without communal education, none of us could have discovered, alone, all the tools that currently extend our physical and mental abilities. Pedagogy and culture make each of us the heir to an extensive chain of human wisdom.

But *Homo sapiens*' dependency on social communication and education is as much of a curse as it is a gift. On the flip side of the coin, it is education's fault that religious myths and fake news propagate so easily in human societies. From the earliest age, our brains trustfully absorb the tales we are told, whether they are true or false. In a social context, our brains lower their guard; we stop acting like budding scientists and become mindless lemmings. This can be good—as when we trust the knowledge of our science teachers, and thus avoid having to replicate every experiment since Galileo's time! But it can also be detrimental, as when we collectively propagate an unreliable piece of "wisdom" inherited from our forebears. It is on this basis that doctors foolishly practiced bloodletting and cupping therapies for centuries, without ever testing their actual impact. (In case you are wondering, both are actually harmful in the vast majority of diseases.)

A famous experiment demonstrates the extent to which social learning can turn intelligent children into unthinking copycats. As early as fourteen months of age, babies readily imitate a person's action, even if it doesn't make sense to them—or perhaps especially when it doesn't.[46] In this experiment, infants see an adult with her hands tied up by a shawl, pressing a button with her head. The infants infer that they can simply press the button with their free hands, and this is how they end up imitating the action, rather than copying it in every detail. If, however, they see the same person pressing a button with her head for no particular reason, hands completely free and perfectly visible, then the babies seem to abandon all reasoning and blindly trust the adult—they faithfully imitate the action with a bow of the head, although this movement is meaningless. The infants' head bow seems to be a precursor of the thousands of arbitrary gestures and conventions that human societies and religions perpetuate. In adulthood, this social conformism persists and grows. Even the most trivial of our perceptual decisions, such as judging the length of a line, are influenced by social context: when our neighbors come to a different conclusion than us, we frequently revise our judgment to align it with theirs, even when their answer seems implausible.[47] In such cases, the social animal in us overrides the rational beast.

In short, our *Homo sapiens* brain is equipped with two modes of learning: an active mode, in which we test hypotheses against the outside world like good scientists, and a receptive mode, in which we absorb what others transmit to us without personally verifying it. The second mode, through a cultural ratchet effect, is what allowed the extraordinary expansion of human societies over the past fifty thousand years. But without the critical thinking that characterizes the first mode, the second becomes vulnerable to the spread of fake news. The active verification of knowledge, the rejection of simple hearsay, and the personal construction of meaning are essential filters to protect us from deceitful legends and gurus. We must therefore find a compromise between our two learning modes: our students must be attentive and confident in their teachers' knowledge, but also autonomous and critical thinkers, actors of their own learning.

We are now touching the second pillar of learning: active engagement.

CHAPTER 8

Active Engagement

TAKE TWO KITTENS. PUT A COLLAR AND LEASH ON THE FIRST ONE. PLACE
the second one in a harness. Finally, connect them to a merry-go-round appa-
ratus which ensures that the movements of the two kittens are strictly linked.
The idea is that the two animals receive identical visual inputs, but one is
active while the other is passive. The former explores the environment on its
own, while the latter moves in exactly the same way, but without control.

This is the classic carousel experiment that Richard Held (1922–2016)
and Alan Hein performed in 1963—a time when the ethics of animal experi-
mentation was clearly not as developed as it is today! This very simple experiment led to an important discovery: active exploration of the world is essential
for the proper development of vision. Over a period of a few weeks, for three
hours a day, the two kittens lived in a large cylinder lined with vertical bars.
Although their visual inputs were very similar, they developed dramatically
different visual systems.[1] Despite the impoverished environment consisting
only of vertical bars, the active kitten developed normal vision. The passive
kitten, on the other hand, lost its visual abilities and, at the end of the experiment, failed basic visual exploration tests. In the cliff test, for example, the
animal was placed on a bridge that it could leave either on the side of a high

cliff or on the shallower side. A normal animal does not hesitate for a second and jumps to the easy side. The passive animal, however, chose at random. Other tests showed that the passive animal failed to develop a proper model of visual space and did not feel out its environment with its paws like normal cats do.

A PASSIVE ORGANISM DOES NOT LEARN

Held and Hein's carousel experiment is the metaphor for our second pillar of learning: active engagement. Converging results from diverse fields suggest that a passive organism learns little or nothing. Efficient learning means refusing passivity, engaging, exploring, and actively generating hypotheses and testing them on the outside world.

To learn, our brain must first form a hypothetical mental model of the outside world, which it then projects onto its environment and puts to a test by comparing its predictions to what it receives from the senses. This algorithm implies an active, engaged, and attentive posture. Motivation is essential: we learn well only if we have a clear goal and we fully commit to reaching it.

Don't get me wrong: active engagement does not mean that children should be encouraged to fidget in class all day long! I once visited a school where the principal told me, with a certain pride, how he applied my ideas: he had equipped his pupils' desks with pedals so that his students could remain active during math class. . . . He had totally missed my point (and showed me the limits of the carousel experiment metaphor). Being active and engaged does not mean that the body must move. Active engagement takes place in our brains, not our feet. The brain learns efficiently only if it is attentive, focused, and active in generating mental models. To better digest new concepts, active students constantly rephrase them into words or thoughts of their own. Passive or, worse, distracted students will not benefit from any lesson, because their brains do not update their mental models of the world. This has nothing to do with actual motion. Two students could be very still yet differ dramatically in the inner movements of their thoughts: one actively follows the course, while the other disengages and becomes passive or distracted.

Experiments show that we rarely learn by merely accumulating sensory statistics in a passive manner. This can happen, but mainly at the lower levels of our sensory and motor systems. Remember those experiments where a child hears hundreds of syllables, computes the transition probabilities between syllables (such as /bo/ and /t^l/), and ends up detecting the presence of words ("bottle")? This type of implicit learning seems to persist even when infants are asleep.[2] However, it is the exception that proves the rule: in the vast majority of cases, and as soon as learning concerns high-level cognitive properties, such as the explicit memory of word meanings rather than their mere form, learning seems to occur only if the learner pays attention, thinks, anticipates, and puts forth hypotheses at the risk of making mistakes. Without attention, effort, and in-depth reflection, the lesson fades away, without leaving much of a trace in the brain.

DEEPER PROCESSING, BETTER LEARNING

Let's take a classical example from cognitive psychology: the effect of word processing depth. Imagine that I present a list of sixty words to three groups of students. I ask the first group to decide whether the words' letters are upper- or lowercase; the second group, whether the words rhyme with "chair"; and the third, whether they are animal names or not. Once the students are finished, I give them a memory test. Which group remembers the words best? Memory turns out to be much better in the third group, who processed the words in depth, at the meaning level (75 percent success), than in the other two groups, who processed the more superficial sensory aspects of the words, either at the letter level (33 percent success) or the rhyme level (52 percent success).[3] We do find a weak implicit and unconscious trace of the words in all groups: learning leaves its subliminal mark within the spelling and phonological systems. However, only in-depth semantic processing guarantees explicit,

detailed memory of the words. The same phenomenon occurs at the level of
sentences: students who make the effort to understand sentences on their own,
without teacher guidance, show much better retention of information.[4] This
is a general rule, which the American psychologist Henry Roediger states as
follows: "Making learning conditions more difficult, thus requiring students
to engage more cognitive effort, often leads to enhanced retention."[5]

Brain imaging is beginning to clarify the origins of this processing depth
effect.[6] Deeper processing leaves a stronger mark in memory because it activates areas of the prefrontal cortex that are associated with conscious word
processing and because these areas form powerful loops with the hippocampus, which stores information in the form of explicit episodic memories.

In the cult film *La Jetée* (1962), by French director Chris Marker
(1921–2012), a voice-over states the following aphorism, which sounds like a
profound truth: "Nothing distinguishes memories from ordinary moments:
only later do they make themselves known, from their scars." A beautiful
adage . . . but a false proverb, because brain imaging shows that at the onset of
memory encoding, the events of our life which will remain engraved in our
memory can already be distinguished from those that will leave no trace: the
former have been processed at a deeper level.[7] By scanning a person while
she is merely exposed to a list of words and images, we can predict which of
those individual stimuli will be later forgotten and which will be retained.
The key predictor is whether they induced activity in the frontal cortex, the
hippocampus, and the neighboring regions of the parahippocampal cortex.
The active engagement of these regions is a direct reflection of the depth to
which these words and images traveled in the brain, and it predicts the strength
of the trace that they leave in memory. An unconscious image enters sensory
areas but creates only a modest wave of activity in the prefrontal cortex. Attention, concentration, processing depth, and conscious awareness transform this
small wave into a neuronal tsunami that invades the prefrontal cortex and
maximizes subsequent memorization.[8]

The role of active engagement and processing depth is confirmed by converging evidence from pedagogical studies in a school context—for example,

learning physics at the undergraduate level. Students must learn the abstract concepts of angular momentum and motor torque. We divide the students into two groups: one group is given ten minutes to experiment with a bicycle wheel, and the other group, ten minutes of verbal explanation and observation of other students. The result is clear: learning is much better in the group that benefited from active interaction with the physical object.[9] Making a course deeper and more engaging facilitates the subsequent retention of information.

This conclusion receives support from a recent review of more than two hundred pedagogical studies in undergraduate STEM courses: traditional lecturing, where students remain passive while the teacher preaches for fifty minutes, is inefficient.[10] Compared with teaching methods that promote active engagement, lecturing systematically yields lower performances. In all disciplines, from math to psychology, biology to computer science, an active student succeeds more. With active engagement, examination scores progress by half a standard deviation, which is considerable, and the failure rate decreases by over 10 percent. But what are the strategies that engage students the most? There is no single miraculous method, but rather a whole range of approaches that force students to think for themselves, such as practical activities, discussions in which everyone takes part, small group work, or teachers who interrupt their class to ask a difficult question and let the students think about it for a while. All solutions that force students to give up the comfort of passivity are effective.

THE FAILURE OF DISCOVERY-BASED TEACHING

None of this is new, you may be thinking, and many teachers already apply these ideas. However, in the pedagogical domain, neither tradition nor intuition can be trusted: we need to scientifically verify which pedagogies actually improve students' comprehension and retention, and which do not. And this is an opportunity for me to clarify a very important distinction. The fundamentally correct view that children must be attentively and actively engaged in their own learning must not be confused with classical constructivism or

discovery learning methods—which are seductive ideas whose ineffective-
ness has, unfortunately, been repeatedly demonstrated.[11] This is a key distinction, but it is rarely understood, in part because the latter pedagogies are also
known as active pedagogies, which is a great source of confusion.

When we talk about discovery learning, what do we mean? This nebula
of pedagogical views can be traced back to Jean-Jacques Rousseau and has
reached us through famous educators such as John Dewey (1859–1952), Ovide
Decroly (1871–1932), Célestin Freinet (1896–1966), Maria Montessori, and,
more recently, Jean Piaget and Seymour Papert (1928–2016). "Do I dare set
forth here," writes Rousseau in *Emile, or On Education*, "the most important, the
most useful rule of all education? It is not to save time, but to squander it." For
Rousseau and his successors, it is always better to let children discover for
themselves and build their own knowledge, even if it implies that they might
waste hours tinkering and exploring. . . . This time is never lost, Rousseau
believed, because it eventually yields autonomous minds, capable not only of
thinking for themselves but also of solving real problems, rather than passively receiving knowledge and spitting out rote and ready-made solutions.
"Teach your student to observe the phenomena of nature," says Rousseau,
"and you will soon rouse his curiosity; but if you want his curiosity to grow,
do not be in too great a hurry to satisfy it. Lay the problems before him and
let him solve them himself."

The theory is attractive. . . . Unfortunately, multiple studies, spread over
several decades, demonstrate that its pedagogical value is close to zero—and
this finding has been replicated so often that one researcher entitled his
review paper "Should There Be a Three-Strikes Rule against Pure Discovery
Learning?" When children are left to themselves, they have great difficulty
discovering the abstract rules that govern a domain, and they learn much less,
if anything at all. Should we be surprised by this? How could we imagine that
children would rediscover, in a few hours and without any external guidance,
what humanity took centuries to discern? At any rate, the failures are resound-
ing in all areas:

- In reading: Mere exposure to written words usually leads to nothing unless children are explicitly told about the presence of letters and their correspondence with speech sounds. Few children manage to correlate written and spoken language by themselves. Imagine the intellectual powers that our young Champollion would need in order to discover that all words beginning with the sound /R/ also bear the mark "R" or "r" at their leftmost end. . . . The task would be out of reach if teachers did not carefully guide children through an ordered set of well-chosen examples, simple words, and isolated letters.

- In mathematics: It is said that at the age of seven, the brilliant mathematician Carl Gauss (1777–1855) discovered, all by himself, how to quickly add the numbers from one to one hundred (think about it—I give the solution in the notes[12]). What worked for Gauss, however, may not apply to other children. Research is clear on this point: learning works best when math teachers first go through an example, in some detail, before letting their students tackle similar problems on their own. Even if children are bright enough to discover the solution by themselves, they later end up performing worse than other children who were first shown how to solve a problem before being left to their own means.

- In computer science: In his book *Mindstorms* (1980), computer scientist Seymour Papert explains why he invented the Logo computer language (famous for its computerized turtle that draws patterns on the screen). Papert's idea was to let children explore computers on their own, without instruction, by getting hands-on experience. Yet the experiment was a failure: after a few months, the children could write only small, simple programs. The abstract concepts of computer science eluded them, and on a problem-solving test, they did no better than untrained children: the little computer literacy they had learned had not spread to other areas. Research shows that explicit teaching, with alternating periods of explanation and hands-on testing, allows children to develop a much deeper understanding of the Logo language and computer science.

I directly experienced the birth of the personal home computer—I was fifteen years old when my father bought us a Tandy TRS-80 with sixteen kilobytes of memory and 48-by-128-pixel graphics. Like others of my generation, I learned to code in the programming language BASIC without a teacher or a class—although I was not alone: my brother and I devoured all the magazines, books, and examples we could get our hands on. I eventually became a reasonably effective programmer . . . but when I entered a master's program in computer science, I became aware of the enormity of my shortcomings: I had been tinkering all this time without understanding the deep, logical structure of programs, nor the proper practices that made them clear and legible. And this is perhaps the worst effect of discovery learning: it leaves students under the illusion that they have mastered a certain topic, without ever giving them the means to access the deeper concepts of a discipline.

In summary, while it is crucial for students to be motivated, active, and engaged, this does not mean that they should be left to their own devices. The failure of constructivism shows that explicit pedagogical guidance is essential. Teachers must provide their students with a structured learning environment designed to progressively guide them to the top as quickly as possible. The most efficient teaching strategies are those that induce students to be actively engaged while providing them with a thoughtful pedagogical progression that is closely channeled by the teacher. In the words of psychologist Richard Mayer, who reviewed this field, the best success is achieved by "methods of instruction that involve cognitive activity rather than behavioral activity, instructional guidance rather than pure discovery, and curricular focus rather than unstructured exploration."[13] Successful teachers provide a clear and rigorous sequence that begins with the basics. They constantly assess their students' mastery and let them build a pyramid of meaning.

And this is indeed what most schools inspired by Montessori do today: they do not let children "marinate" without doing anything; instead, they propose a whole series of rational and hierarchical activities, whose purpose is first carefully demonstrated by teachers before being carried out independently by children. Active engagement, pleasure, and autonomy, under the guidance of an explicit teaching method and with stimulating pedagogical materials: these are the ingredients for a winning recipe whose effectiveness has been repeatedly demonstrated.

Pure discovery learning, the idea that children can teach themselves, is one of many educational myths that have been debunked but still remain curiously popular. It belongs to a collection of urban legends that mar the educational field, and at least two other major misconceptions are linked to it:[14]

- The myth of the digital native: Children of the new generation, unlike their parents, have been bathed in computers and electronics since their earliest years. As a result, according to this myth, these native *Homo zappiens* are champions of the digital world, for whom bits and bytes are completely transparent, and who surf and switch between digital media with incredible ease. Nothing could be further from the truth: research shows that these children's mastery of technology is often superficial, and that they are just as bad as any of us at multitasking. (As we have seen, the central bottleneck that prevents us from doing two things at once is a fundamental property of our brain architecture, present in all of us.)
- The myth of learning styles: According to this idea, each student has his or her own preferred learning style—some are primarily visual learners, others auditory, yet others learn better from hands-on experience, and so on. Education should therefore be tailored to each student's favorite mode of knowledge acquisition. This is also patently false:[15] as amazing as it may seem, there is no research supporting the notion that children differ radically in their preferred learning modality. What is true is that some teaching strategies work better than others—but when they do, this superiority applies to all of us, not just a subgroup. For instance, experiments show that all of us have an easier time remembering a picture than a spoken word, and that our memory is even better when the information is conveyed by both modalities—an audiovisual experience. Again, this is the case for all children. There is simply no evidence in favor of the existence of subtypes

186   How We Learn

of children with radically different learning styles, such that type A children learn better with strategy A, and type B children with strategy B. For all we know, all humans share the same learning algorithm.

What about all the special education books and software that claim to tailor education to each child's needs? Are they worthless? Not necessarily. Children do vary dramatically, not in learning style, but in the speed, ease, and motivation with which they learn. In first grade, for instance, the top 10 percent of children already read more than four million words per year, whereas the bottom 10 percent read less than sixty thousand[16]—and dyslexic children may not read at all. Developmental deficits such as dyslexia and dyscalculia may come in several varieties, and it is often useful to carefully diagnose the exact nature of the impairment in order to adapt the lessons. Children do benefit from pedagogical interventions whose contents are tailored to their specific difficulties. For instance, many children, even in advanced mathematics, fail to understand how fractions work—in this case, the teacher should shed the current curriculum and return to the basics of numbers and arithmetic. However, every teacher should also keep in mind that all children learn using the same basic machinery—one that prefers focused attention to dual tasking, active engagement to passive lecturing, detailed error correction to phony praise, and explicit teaching over constructivism or discovery learning.

CURIOSITY AND HOW TO PIQUE IT

> All men by nature desire to know.
>
> Aristotle, *Metaphysics* (c. 335 BCE)

> I have no special talent. I am only passionately curious.
>
> Albert Einstein (1952)

One of the foundations of active engagement is curiosity—the desire to learn, or the thirst for knowledge. Piquing children's curiosity is half the battle. Once their attention is mobilized and their mind in search of an explanation, all that is left to do is guide them. Starting in kindergarten, the most curious students are also those who do better in reading and math.[17] Keeping children curious is therefore one of the key factors for successful education. But what exactly is curiosity? To what Darwinian necessity does it respond, and to what kind of algorithm does it correspond?

Rousseau wrote in *Emile, or On Education*, "One is curious only to the extent that one is educated." Here again, he was wrong: curiosity is not an effect of instruction, a function that we must acquire. It is already present at an early age and is an integral part of our human brain circuitry, a key ingredient of our learning algorithm. We do not simply passively wait for new information to reach us—as do, foolishly, most current artificial neural networks, which are simple input-output functions passively submitted to their environment. As Aristotle noted, we humans are born with a passion to know, and we constantly seek novelty, actively exploring our environment to discover things we can learn.

Curiosity is a fundamental drive of the organism: a propulsive force that pushes us to act, just like hunger, thirst, the need for security, or the desire to reproduce. What role does it play in survival? It is in the interest of most animal species (mammals, but also many birds and fish) to explore their environment in order to better monitor it. It would be risky to set up a nest, lair, burrow, den, hole, or home without checking the surroundings. In an unstable universe populated by predators, curiosity can make all the difference between life and death—and this is why most animals regularly pay security visits to their territory, carefully checking for anything unusual and investigating novel sounds or sights. . . . Curiosity is the determination that pushes animals out of their comfort zones in order to acquire knowledge. In an uncertain world, the value of information is high and must ultimately be paid in Darwin's own currency: survival.

Curiosity is therefore a force that encourages us to explore. From this perspective, it resembles the drive for food or sexual partners, except that it is motivated by an intangible value: the acquisition of information. Indeed, neurobiological studies show that, in our brains, the discovery of previously unknown information brings its own reward: it activates the dopamine circuit. Remember, this is the circuit that fires in response to food, drugs, and sex. In primates, and probably in all mammals, this circuit responds not solely to material rewards, but also to new information. Some dopaminergic neurons signal a future information gain, as if the anticipation of novel information brings its own gratification.[18] Thanks to this mechanism, rats can be conditioned not only to food or drugs, but also to novelty: they quickly develop a preference for places that contain new objects and thereby satisfy their curiosity, as opposed to dull places where nothing ever happens.[19] We do not act any differently when we move to a big city for a change of scenery or when, eager for the latest gossip, we frantically scroll through Facebook or Twitter.

Humans' appetite for knowledge passes through the dopamine circuit even when it involves a purely intellectual curiosity. Imagine lying in an MRI and being asked Trivial Pursuit questions, such as, "Who was the president of the United States when Uncle Sam first got his beard?"[20] For each question, before satisfying your curiosity, the experimenter asks how eager you are to know the answer. What are the neuronal correlates of this subjective feeling of being curious? The degree of curiosity that you report correlates tightly with the activity of the nucleus accumbens and the ventral tegmental area, two essential regions of the dopamine brain circuit. The more curious you are, the more these regions light up. Their signals arise in anticipation of the answer: even before your curiosity is satisfied, the simple fact of knowing that you will soon know the answer excites your dopaminergic circuits. Expectation of a positive event brings its own reward.

These curiosity signals are obviously useful, because they predict how much you learn. Memory and curiosity are linked—the more curious you are about something, the more likely you are to remember it. Curiosity even transfers to nearby events: when your curiosity is heightened, you remember incidental details such as the face of a passerby or the person who taught you the information that you were so eager to learn. The degree of craving for knowledge controls the strength of memory.

Through the dopamine circuit, the satisfaction of our appetite to learn—or even the mere anticipation of that satisfaction—is deeply rewarding. Learning possesses intrinsic value for the nervous system. What we call curiosity is nothing more than the exploitation of this value. As such, our species is probably special because of its unmatched ability to learn. As hominization progressed, our ability to represent the world progressed. We are the only animals who formulate formal theories of the world in a language of thought. Science has become our ecological niche: *Homo sapiens* is the only species without a specific habitat, because we learn to adapt to any environment.

Mirroring the extraordinary expansion of our learning abilities, human curiosity seems to have increased tenfold. Over the course of our evolution, we have acquired an extended form of curiosity, called "epistemic curiosity": the pure desire for knowledge in all fields, including the most abstract. Like other mammals, we play and explore—not only through real movement, but also through thought experiments. Whereas other animals visit the space around them, we explore conceptual worlds. Our species also experiences specific epistemic emotions that guide our thirst for knowledge. We rejoice, for example, in the symmetry and pure beauty of mathematical patterns: a clever theorem can move us much more than a piece of chocolate.

Mirth seems to be one of those uniquely human emotions that guide learning. Our brain triggers a mirth reaction when we suddenly discover that one of our implicit assumptions is wrong, forcing us to drastically revise our mental model. According to the philosopher Dan Dennett, hilarity is a contagious social response that spreads as we draw each other's attention to an unexpected piece of information.[21] And, indeed, all things being equal, laughing during learning seems to increase curiosity and enhance subsequent memory.[22]

WANTING TO KNOW: THE SOURCE OF MOTIVATION

Several psychologists have tried to specify the algorithm that underlies human curiosity. Indeed, if we understood it better, we could perhaps gain control over this essential ingredient of our learning scheme, and even reproduce it in a machine that would eventually imitate the performance of the human species: a curious robot.

This algorithmic approach is beginning to bear fruit. The greatest psychologists, from William James to Jean Piaget to Donald Hebb, have speculated on the nature of the mental operations that underlie curiosity. According to them, curiosity is the direct manifestation of children's motivation to understand the world and build a model of it.[23] Curiosity occurs whenever our brains detect a gap between what we already know and what we would like to know—a potential learning area. At any given moment, we choose, from the various actions that are accessible to us, those that are most likely to reduce this knowledge gap and acquire useful information. According to this theory, curiosity resembles a cybernetic system that regulates learning, similar to the famous Watt governor, which opens or closes the throttle valve on a steam engine in order to regulate steam pressure and maintain a fixed speed. Curiosity would be the brain's governor, a regulator that seeks to maintain a certain learning pressure. Curiosity guides us to what we think we can learn. Its opposite, boredom, turns us away from what we already know, or from areas that, according to our past experience, are unlikely to have anything left to teach us.

This theory explains why curiosity is not directly related to the degree of surprise or novelty but instead follows a bell curve.[24] We have no curiosity for the unsurprising—things that we have seen a thousand times before are boring. But we are also not attracted to things that are too novel or surprising, or so confusing that their structure eludes us—their very complexity deters us. Between the boredom of the too simple and the repulsion of the too complex, our curiosity naturally directs us toward new and accessible fields. But this attraction keeps changing. As we master them, the objects that once seemed attractive lose their appeal, and we redirect our curiosity toward new challenges. This is why babies initially seem so passionate about the most trivial things: grasping their toes, closing their eyes, playing peekaboo. . . . Everything is new to them and is a potential source of learning. Once they squeeze out all the knowledge that can be gained from those experiments, they lose interest—for exactly the same reason that no scientist reproduces Galileo's experiments anymore: what is known becomes boring.

The same algorithm also explains why we sometimes turn away from an area that once seemed attractive but proved to be too difficult. Our brain evaluates the speed of learning, and curiosity is turned off if our brain detects that we are not progressing fast enough. We all know of children who, say, return from a concert with a passion for the violin . . . only to give it up after a few weeks, when they realize that mastery of the instrument does not come easily. Those who keep playing either set more modest goals (e.g., play a little better every day) or, if they truly aim to become professional musicians, sustain their motivation through parental and social support and constant reminders of their long-term goals.

Two French engineers, Frédéric Kaplan and Pierre-Yves Oudeyer, have implemented curiosity in a robot.[25] Their algorithm includes several modules. The first is a classic artificial learning system that constantly tries to predict the state of the outside world. The second, more innovative module evaluates the performance of the first: it measures the recent learning speed and uses it to predict the areas in which the robot will learn the most. The third ingredient is a reward circuit that places greater value on actions that are predicted to lead to more efficient learning. As a result, the system naturally focuses on those areas where it believes that it will learn the most, which is the very definition of curiosity, according to Kaplan and Oudeyer.

When their curious robot, equipped with this algorithm, is placed on a baby mat, it behaves exactly like a young child. For a few minutes, it becomes enthused about a particular object and spends all its time, for example, repeatedly lifting a stuffed elephant ear. As it progressively learns all there is to know about an item, its curiosity dwindles. At one point, it turns away and actively seeks another source of stimulation. After an hour, it stops exploring the mat: a digital form of boredom sets in as the robot comes to believe that everything that could be learned is now known.

The analogy with a small child is striking. Even babies a few months old

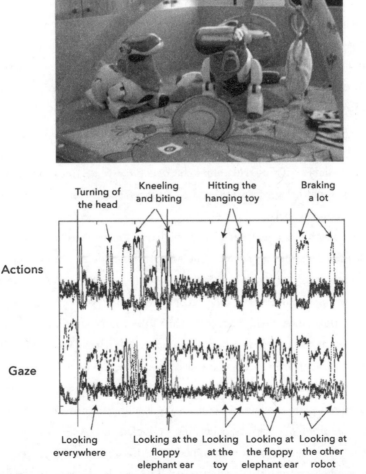

Curiosity is an essential ingredient of our learning algorithm, which is only beginning to be reproduced in machines. Here, a small robot explores a play mat. Curiosity is implemented by a reward function that favors the choice of whichever action maximizes the potential to learn. As a consequence, the robot successively tries out each toy on the mat and each action at its disposal. Once it masters one aspect of the world, it loses interest and redirects its attention elsewhere.

orient toward stimuli of intermediate complexity, neither too simple nor too complex, but whose structure is just right to be quickly learnable. (This trait of infants' curiosity has been described as the "Goldilocks effect."[26]) To maximize what they learn, we have to constantly enrich their environment with new objects that are just stimulating enough to not be discouraging. It is adults' responsibility to provide them with a well-designed pedagogical hierarchy that progressively takes them to the top, constantly stimulating their drive for knowledge and novelty.

This vision of curiosity leads to an interesting prediction. It implies that in order for children to be curious, they must be aware of what they do not yet know. In other words, they must possess *metacognitive* faculties at an early age. "Metacognition" is cognition over cognition: the set of higher-order cognitive systems that monitor our mental processes. According to the gap theory of curiosity, metacognitive systems must constantly supervise our learning, evaluating what we know and don't know, whether we are wrong or not, whether we are fast or slow, and so on and so forth—metacognition encompasses everything we know about our own minds.

Metacognition plays a key role in curiosity. Indeed, to be curious is to want to know, and that implies knowing what you don't already know. And once again, recent experiments confirm that from the age of one and perhaps even earlier, children understand that there are things they do not know.[27] Indeed, babies of that age readily turn to their caregiver whenever they are unable to solve a problem alone. Knowing that they don't know leads them to ask for more information. This is the early manifestation of epistemic curiosity: the irresistible desire to know.

THREE WAYS SCHOOL CAN KILL CURIOSITY

All parents are nostalgic for the days when their toddler was filled with curiosity. Between ages two and five, children are curious about everything. Their favorite word is often *why*: they never stop experimenting on the world and questioning adults in order to quench their thirst for knowledge. Surprisingly, however, this appetite, which seems insatiable, eventually dies out, often after a few years of school. Some children remain curious about everything, but many close themselves off to such intrigue. Their active engagement turns into dull passivity. Can the science of curiosity explain why? We do not yet have all the answers, but I would like to propose a few hypotheses.

First, children may lose their curiosity because they lack cognitive stimulation tailored to their needs. According to the algorithm we have just described, it is entirely normal for curiosity to dwindle over time. As learning progresses, the expected learning gain shrinks: the better we master a field, the more we reach the limits of what it can offer, and the less interested we are in it. To maintain curiosity, schools must therefore continually provide children's supercomputing brains with stimulants that match their intelligence. This is not always the case. In a standard classroom, the most advanced students often lack stimulation: after a few months, their curiosity fades and they no longer expect much from school, because their metacognitive system has learned that, unfortunately, they are unlikely to learn much more.

At the other end of the spectrum, students who struggle in school may wither away for the opposite reason. Metacognition remains the main culprit: after a while, they no longer have any reason to be curious, because they have learned . . . that they do not succeed in learning. Their past experience has engraved a simple (though false) rule in the depths of their metacognitive circuits: I am incapable of learning such and such topic (math, reading, history, whatever). Such dismay is not uncommon: many girls convince themselves that mathematics is not for them,[28] and children from underprivileged neighborhoods sometimes come to believe that school is hostile for them and teaches nothing useful for their future. Such metacognitive judgments are disastrous because they demotivate students and nip their curiosity in the bud.

The solution is to boost these children's confidence back up, step by step, by showing them that they are perfectly capable of learning, provided the problems are adapted to their level, and that learning brings its own reward. The theory of curiosity says that when children are discouraged, whether they are far ahead or far behind at school, what matters most is to restore their desire to learn by offering them stimulating problems carefully tailored to their current level. First, they rediscover the pleasure of learning something new—and then, slowly, their metacognitive system learns that they *can* learn, which puts their curiosity back on track.

Another scenario that can lead to children losing interest is when curiosity is punished. A child's appetite for discovery can be ruined by an overly rigid pedagogical strategy. Teaching through traditional lectures tends to discourage children from participating or even from thinking. It can convince children that they are simply being asked to sit there and remain quiet until the end of class. The neurophysiological interpretation of this situation is simple: within the dopamine circuit, the reward signals induced by curiosity and its satisfaction compete with external rewards and punishments. It is therefore possible to discourage curiosity by punishing each exploration attempt. Picture a child who repeatedly tries to participate and is systematically reprimanded, mocked, or punished: "Silly question. You'd better be quiet or you'll stay an extra half an hour after school. . . ." This child quickly learns to inhibit their curiosity drive and stop participating in class: the curiosity-based reward that the dopamine system expects—the pleasure of learning something new—is largely countered by the direct negative signals that the same circuit receives. Repeated punishment leads to learned helplessness, a kind of physical and mental paralysis associated with stress and anxiety, which has been shown to inhibit learning in animals.[29]

The solution? Most teachers already know it. It is simply a matter of rewarding curiosity instead of punishing it: encouraging questions (however imperfect they may be), asking children to give presentations on subjects they love, rewarding them for taking initiative. . . . The neuroscience of motivation is extremely clear: the desire to do action X must be associated with an expected reward, be it material (food, comfort, social support) or cognitive (acquisition of information). Too many children lose all curiosity because they learn, at their own expense, to expect no reward from school. (Grades, which I will get to shortly, often contribute to this sad state of affairs.)

The third factor that can discourage curiosity is the social transmission of knowledge. Remember that two modes of learning coexist in the human species: the active mode, where children constantly experiment and question themselves like good budding scientists, and the receptive mode, where they simply record what others teach them. School often encourages only the second mode—and it may even discourage the first, if children assume that teachers always know everything better than students do.

Can a teacher's attitude really kill a child's natural curiosity?[30] Sadly, recent experiments suggest that the answer is yes. In her childhood cognition lab at MIT, the American developmental psychologist Laura Schulz presents kindergartners with a strange contraption: a set of plastic tubes hidden in various places that contain all sorts of unexpected toys, such as a mirror, a horn, a game with lights, and a music box. When you give such a gadget to children without saying anything, you immediately set off their curiosity: they explore, rummage, forage, and poke around until they find most of the hidden rewards. Now, take a new group of kindergartners and put them into the passive, receptive pedagogical mode. All you have to do is give them the object while saying, "Look, let me show you my toy. This is what it does . . ." and then play the music box, for instance. One might think that this would stimulate the children's curiosity . . . but it has the opposite effect: exploration massively decreases following this kind of introduction. Children seem to make the (often correct) assumption that the teacher is trying to help them as much as possible, and that he has therefore introduced them to all the interesting functions of the device. In this context, there is no need to search: curiosity is inhibited.

Further experiments show that children take into account the teacher's past behavior. When a teacher always makes exhaustive demonstrations, students lose curiosity. If the teacher demonstrates one of the functions of a new toy, children do not explore all its facets, because they think that the teacher has already explained everything there is to know. If, on the contrary, the teacher gives evidence that he doesn't always know everything, then the children keep searching.

So, what is the right approach? I suggest always keeping the concept of active engagement in mind. Maximally engaging a child's intelligence means constantly feeding them with questions and remarks that stimulate their imagination and make them want to go deeper. It would be out of the question to let students discover everything for themselves—this would be falling back into the trap of discovery-based learning. The ideal scenario is to offer the guidance of a structured pedagogy while encouraging children's creativity by letting them know that there are still a thousand things to discover. I remember a teacher who, just before summer vacation, told me, "You know, I just read a little math problem I couldn't solve. . . ." And this is how I found myself ruminating on this question all summer, trying to do better than the teacher could. . . .

Mustering children's active engagement goes hand in hand with another necessity: tolerating their errors while quickly correcting them. This is our third pillar of learning.

CHAPTER 9

Error Feedback

Everyone should learn to happily make errors.... To think is to move from one error to the next.

Alain, Propos sur l'éducation (1932)

The only man who never makes a mistake is the man who never does anything.

Attributed to Theodore Roosevelt (1900)

IN 1940, THE YOUNG ALEXANDER GROTHENDIECK (1928-2014) WAS ONLY eleven or twelve years old. He did not know that he would become one of the most influential mathematicians of the twentieth century who would inspire a whole generation. (His revolutionary ideas played a major role in the founding, in 1958, of the famous Institut des Hautes Études Scientifiques, in France, which has yielded more than a dozen Fields Medal winners.) But young Alexander was already doing mathematics ... with moderate success. Here is an excerpt from his memoir:

 Around the age of eleven or twelve, while I was detained at the concentration camp of Rieucros (near Mende), I discovered compass tracing games. I was particularly thrilled by the six-branched rosettes that one gets when dividing a circle into six equal parts by turning a compass six times around the circumference and returning right back to the starting point. This experimental observation convinced me that the length of the circumference was exactly six times that of the radius. When later . . . I saw in a textbook that the relationship was supposed to be much more complicated, that we had $L = 2 \pi R$ with $\pi = 3.14 \ldots$, I was convinced that the book was wrong, that its authors . . . must have ignored this very simple tracing exercise that clearly showed that $\pi = 3$.

The confidence that a child can have in his own insight, by trusting his own faculties rather than taking for granted the things he learns at school or reads in a textbook, is a precious thing. Yet this confidence is constantly discouraged.

Many will see in the experience I just reported the example of a childish brashness that later had to bow in front of the received knowledge—the whole situation bordering on the ridiculous. As I experienced this episode, however, there was no sense of disappointment or ridicule, but just the feeling of having made a genuine discovery . . . : that of a mistake.[1]

What an extraordinary confession, and what a lesson in humility, when one of the world's greatest mathematicians admits to making the colossal blunder of believing that pi equals three. . . . Yet, Grothendieck was quite right about one thing: the key role of errors in learning. Making mistakes is the most natural way to learn. The two terms are virtually synonymous, because every error offers an opportunity to learn.

The Shadoks, a French cartoon that was popular when I was a child, whimsically elevated this concept to the rank of a general principle: "Only by continually trying do you end up succeeding.... In other words, the more you fail, the more likely you are to succeed!" And with perfect logic, since the rocket they were trying to launch had only one chance in a million of taking off, the Shadoks hastily ran through the first 999,999 failures in order to finally reach success....

Humor aside, it would be practically impossible to progress if we did not start off by failing. Errors always recede as long as we receive feedback that tells us how to improve. This is why error feedback is the third pillar of learning, and one of the most influential educational parameters: the quality and accuracy of the feedback we receive determines how quickly we learn.[2]

SURPRISE: THE DRIVING FORCE OF LEARNING

Remember the learning algorithms that we discussed in the first chapter, which enabled a hunter to adjust his viewfinder or an artificial neural network to tune its hidden weights? The idea is simple: you first try, even if it means failing, and the size and direction of your error tells you how to improve on the next trial. Thus, the hunter aims, shoots, evaluates how much he missed the target, and uses this error feedback to adjust his next shot. This is how marksmen fine-tune their rifles—and how, on a larger scale, artificial neural networks adjust the millions of parameters that define their internal models of the outside world.

Does the brain work the same way? As early as the 1970s, data started to accumulate in favor of this hypothesis. Two American researchers, Robert Rescorla and Allan Wagner, made the following hypothesis: the brain learns only if it perceives a gap between what it predicts and what it receives. No learning is possible without an error signal: "Organisms only learn when events violate their expectations."[3] In other words, surprise is one of the fundamental drivers of learning.

The Rescorla-Wagner theory nicely explains the details of a learning paradigm called "classical conditioning." Everyone has heard of Pavlov's dog. In Pavlovian conditioning experiments, a dog hears a bell, which is an initially neutral and inefficient stimulus. After repeated pairing with food, however, the same bell ends up triggering a conditioned reflex. The dog salivates whenever he hears the bell, because he has learned that this sound systematically precedes the arrival of food. How does the theory explain these findings? The Rescorla-Wagner rule assumes that the brain uses sensory inputs (the sensations generated by the bell) to predict the probability of a subsequent stimulus (food). It works like this:

- The brain generates a prediction by computing a weighted sum of its sensory inputs.
- It then calculates the difference between this prediction and the actual stimulus it receives: this is the *prediction error*, a fundamental concept of the theory, which measures the degree of surprise associated with each stimulus.
- The brain then uses this surprise signal to correct its internal representation: the internal model changes in direct proportion to both the strength of the stimulus and the value of the prediction error. The rule is such that it guarantees that the next prediction will be closer to reality.

This theory already contains all the seeds of our three pillars of learning: learning occurs only if the brain selects the appropriate sensory inputs (attention), uses them to produce a prediction (active engagement), and evaluates the accuracy of the prediction (error feedback).

The equation that Rescorla and Wagner introduced in 1972 was remarkably prescient. It is practically identical to the "delta rule" that was later used in artificial neural networks—and both are simplified versions of the error backpropagation rule, which is now used in virtually all current supervised learning systems (where the network is given explicit feedback about the response it should have produced). Moreover, in reward-based machine learning (where the network is just told how wrong it is), a similar equation can still be used: the network predicts the reward, and the difference between that prediction and the actual reward is what is used to update the internal representation.

We can therefore affirm that today's silicon-based learning machines rely on equations directly inspired by neuroscience. As we saw above, the human brain goes even further: to extract as much information as possible from each learning episode, it uses a language of thought and statistical models much more refined than those of current neural networks. However, Rescorla and Wagner's basic idea remains correct: the brain tries to predict the inputs it receives and adjusts these predictions according to the degree of surprise, improbability, or error. To learn is to curtail the unpredictable.

Rescorla and Wagner's theory had considerable impact because it represented a major improvement over previous theories based on the concept of associative learning. In the past, the common belief was that the brain merely learned to associate the sound of a bell with food, rather than predicting one from the other. According to this associationist view, the brain records all coincidences between stimuli and responses in a purely passive fashion. However, even for Pavlovian conditioning, this vision is demonstrably false.[4] Even a dog's brain is not a passive organ that simply absorbs associations. Learning is active and depends on the degree of surprise linked to the violation of our expectations.

Forward blocking provides one of the most spectacular refutations of the associationist view.[5] In blocking experiments, an animal is given two sensory clues, say, a bell and a light, both of which predict the imminent arrival of food. The trick is to present them sequentially. We start with the light: the animal learns that whenever the light is on, it predicts the arrival of food. Only then do we introduce dual trials where both light and bell predict food. Finally, we test the effect of the bell alone. Surprise: it has no effect whatsoever! Upon hearing the bell, the animal does not salivate; it seems utterly oblivious to the repeated association between the bell and the food reward. What happened? The finding is incompatible with associationism, but it fits perfectly with the Rescorla-Wagner theory. The key idea is that the acquisition of the first association (light and food) blocked the second one (bell and food). Why? Because the prediction based on light alone suffices to explain everything. The animal already knows that the light predicts the food, so its brain does not generate any prediction error during the second part of the test, where the light and the bell together predict the food. Zero error, zero learning—and thus, the dog does not acquire any knowledge of the association between the sound and the food. Whichever rule is learned first blocks the learning of the second.

This forward blocking experiment clearly demonstrates that learning does not work by association. After all, the bell-food pairing was repeated hundreds of times, yet it failed to induce any learning. The experiment also shows that no learning occurs in the absence of surprise: a prediction error is essential to learning—at least in dogs. And growing evidence suggests that prediction-error systems are present in the brains of all sorts of species.

It is important to note that the error signal we are talking about is an *internal* signal that travels in the brain. We do not need to make an actual error in order to learn—all we need is a discrepancy between what we expected and what we got. Consider a simple binary choice—say, whether Pablo Picasso's second name is Diego or Rodrigo. Suppose that I am lucky enough to venture a correct guess on the first try (by saying Diego—his full name is actually Pablo Diego José Francisco de Paula Juan Nepomuceno María de los Remedios Cipriano de la Santísima Trinidad Ruiz y Picasso!). Do I learn anything? Of course. Even though I answered correctly on the first try, my confidence was low. By chance alone, I had only a fifty-fifty chance of being right. Because I was uncertain, the feedback I received did provide new information: it assured me that my randomly chosen answer was actually 100 percent right. According to the Rescorla-Wagner rule, this new information generates an error signal that measures the gap between what I predicted (a 50 percent chance of being right) and what I now know (a 100 percent certainty of knowing the right answer). In my brain, this error signal spreads and updates my knowledge, thus increasing my chances of responding "Diego" the next time I'm asked. It would be wrong, therefore, to believe that what matters for learning is to make a lot of mistakes, like the Shadoks hastily failing their first

999,999 rocket launches! What matters is receiving explicit feedback that reduces the learner's uncertainty.

No surprise, no learning: this basic rule now seems to have been validated in all kinds of organisms—including young children. Remember that surprise is one of the basic indicators of babies' early skills: they stare longer at any display that magically presents them with surprising events that violate the laws of physics, arithmetic, probability, or psychology (see figure on page 55 and figure 5 in the color insert). But children do not just stare every time they are surprised: they demonstrably learn.

To reach this conclusion, American psychologist Lisa Feigenson performed a series of experiments showing that whenever children perceive an event as impossible or improbable, learning is triggered.[6] For instance, when babies see an object mysteriously passing through a wall, they stare at this impossible scene . . . and subsequently remember better the sound that the object made, or even the verb that an adult used to describe the action ("Look, I just bleeked the toy."). If we give the babies this object, they play with it for much longer than they do with a similar toy that did not violate the laws of physics. Their seemingly playful behavior actually shows that they are actively trying to understand what happened. As scientists in the crib, they perform experiments in an attempt to replicate what they saw. For example, if the object just passed through a wall, they hit it, as if to test its solidity; whereas if they saw it violate the laws of gravity and remain mysteriously suspended in midair, they make it fall from a table, as if to check its levitation powers. In other words, the nature of the unpredictable scene that they observe determines how they later act to adjust their hypotheses. This is exactly what the theory of error backpropagation predicts: every unexpected event leads to corresponding adjustment of the internal model of the world.

All these phenomena have been documented in eleven-month-old babies, but they are probably present at a much earlier age. Learning by error correction is universally widespread in the animal world, and there is every reason to believe that error signals govern learning from the very beginning of life.

THE BRAIN SWARMS WITH ERROR MESSAGES

Error signals play such a fundamental role in learning that virtually all brain areas can be shown to transmit error messages (see figure 17 in the color insert).[7] Let's start with an elementary example: Imagine hearing a series of identical notes, A A A A A. Each note elicits a response in the auditory areas of your brain—but as the notes repeat, those responses progressively decrease. This is called "adaptation," a deceptively simple phenomenon that shows that your brain is learning to predict the next event. Suddenly, the note changes: A A A A A#. Your primary auditory cortex immediately shows a strong surprise reaction: not only does the adaptation fade away, but additional neurons begin to vigorously fire in response to the unexpected sound. And it is not just repetition that leads to adaptation: what matters is whether the notes are predictable. For instance, if you hear an alternating set of notes, such as A B A B A, your brain gets used to this alternation, and the activity in your auditory areas again decreases. This time, however, it is an unexpected repetition, such as A B A B B, that triggers a surprise response.[8]

The auditory cortex seems to perform a simple calculation: it uses the recent past to predict the future. As soon as a note or a group of notes repeats, this region concludes that it will continue to do so in the future. This is useful because it keeps us from paying too much attention to boring, predictable signals. Any sound that repeats is squashed at the input side, because its incoming activity is canceled by an accurate prediction. As long as the input sensory signal matches the prediction that the brain generates, the difference is zero, and no error signal gets propagated to higher-level brain regions. Subtracting the prediction shuts down the incoming inputs—but only as long as they are predictable. Any sound that violates our brain's expectations, on the contrary, is amplified. Thus, the simple circuit of the auditory cortex acts as a filter: it transmits to the higher levels of the cortex only the surprising and unpredictable information which it cannot explain by itself.

Whatever input a brain region cannot explain is therefore passed on to the next level, which then attempts to make sense of it. We may conceive of the cortex as a massive hierarchy of predictive systems, each of which tries to explain the inputs and exchanges the remaining error messages with the others, in the hope that they may do a better job.

For instance, hearing the sequence C C G generates a low-level error signal in the auditory cortex, because the final G differs from the previous notes. Higher-level regions, however, may recognize the whole sequence as a known melody (the beginning of "Twinkle, Twinkle, Little Star"). The surprise caused by the final G is thus only transient: it is quickly explained by a higher-level representation of the whole melody, and the surprise signal stops there—the G, although new, does not generate any surprise in the inferior prefrontal cortex, which can encode entire musical phrases. On the other hand, the repetition of C C C may have the opposite effect: because it is monotonous, it does not generate any error signal in early auditory areas, but it creates surprise in higher-level areas that code for the melody, which predicted a rise to G rather than yet another C. Here, the surprise is that there is no surprise! Even macaque monkeys, like us humans, possess these two levels of auditory processing: the local processing of individual notes in the auditory cortex and the global representation of the melody in the prefrontal cortex.[9]

Error signals such as these seem to be present in every region of the brain. All over the cortex, neurons adapt to repeated and predictable events, and react with an increased discharge whenever a surprising event occurs. The only thing that changes from one brain area to the next is the type of violation that can be detected. In the visual cortex, the presentation of an unexpected image is what triggers a surge of activity.[10] Language areas, for their part, react to abnormal words within a sentence. Take, for example, the following sentence:

"I prefer to eat with a fork and a camel."

Your brain has just generated an N400 wave, an error signal evoked by a word or an image which is incompatible with the preceding context.[11] As its name suggests, this is a negative response that occurs at about four hundred milliseconds after the anomaly and arises from neuronal populations of the left temporal cortex that are sensitive to word meaning. On the other hand, Broca's area in the inferior prefrontal cortex reacts to errors of syntax, when the brain predicts a certain category of word and receives another,[12] as in the following sentence:

"Don't hesitate to take your whenever medication you feel sick."

This time, just after the unexpected word "whenever," the areas of your brain that specialize in syntax emitted a negative wave immediately followed by a P600 wave—a positive peak that occurs around six hundred milliseconds. This response indicates that your brain detected a grammar error and is trying to repair it.

The brain circuit in which predictive and error signals have been best demonstrated is the reward circuit.[13] The dopamine network not only responds to actual rewards, but it also constantly anticipates them. Dopaminergic neurons located in a small nucleus of cells called the "ventral tegmental area" do not simply respond to the pleasures of sex, food, or drink; they actually signal the difference between the expected reward and the one that was obtained, i.e., the prediction error. So, if an animal receives a reward without any warning, say, an unexpected drop of sugar water, this pleasant surprise results in neuronal firing. But if this reward is preceded by a signal that predicts it, then the same sweet syrup no longer causes any reaction. It is now the signal itself that causes a surge of activity in dopamine neurons: learning shifts the response closer to the signal that predicts the reward.

Thanks to this predictive learning mechanism, arbitrary signals can become the bearers of reward and trigger a dopamine response. This secondary reward effect has been demonstrated with money in humans and with the mere sight of a syringe in drug addicts. In both cases, the brain anticipates future rewards. As we saw in the first chapter, such a predictive signal is extremely useful for learning, because it allows the system to criticize itself and to foresee the success or failure of an action without having to wait for external confirmation. This is why actor-critic architectures, in which one neural network learns to criticize the actions of another, are now universally used in artificial intelligence to solve the most complex problems, such as learning to play the game of Go. Generating a prediction, detecting one's error, and correcting oneself are the very foundations of effective learning.

ERROR FEEDBACK IS NOT SYNONYMOUS WITH PUNISHMENT

> I have often been struck by the fact that science teachers, even more than other teachers, cannot understand that their students may not understand. Very few of them have delved deeply into the topics of error, ignorance, and thoughtlessness.
>
> Gaston Bachelard, *The Formation of the Scientific Mind* (1938)

How can we make the most of the error signals that our neurons constantly exchange? For a child or an adult to learn effectively, their environment (be it parents, school, university . . . or just a video game) must provide them with quick and accurate feedback. Learning is faster and easier when students receive detailed error feedback that tells them precisely where they stumbled and what they should have done instead. By providing rapid and precise feedback on errors, teachers can considerably enrich the information available to their students to correct themselves. In artificial intelligence, this type of learning, known as "supervised," is the most effective, because it allows the machine to quickly identify the source of failure and to amend itself.

It is crucial to understand, however, that such error feedback has nothing to do with punishment. We do not punish an artificial neural network; we simply tell it about the responses that it got wrong. We provide it with a maximally informative signal that notifies it, bit by bit, of the nature and sign of its errors.

In this respect, computer science and pedagogy truly see eye to eye. Indeed, the meta-analyses conducted by Australian education specialist John

Hattie clearly show that the quality of the feedback that students receive is one of the determinants of their academic success.[14] Setting a clear goal for learning and allowing students to approach it gradually, without dramatizing their inevitable mistakes, are the keys to success.

Good teachers are already well aware of these ideas. Every day, they witness the Roman dictum *errare humanum est*: to err is human. With a compassionate eye, they look kindly upon their students' mistakes, because they realize that no one learns without making errors. They know that they should diagnose, as dispassionately as possible, the exact areas of difficulty for their students and help them find the best solutions. With experience, these teachers build up a catalog of errors, because all students repeatedly fall into the same old traps. These teachers find the right words to console, reassure, and restore the self-confidence of their students, all the while allowing them to amend their erroneous mental representations. They are here to tell the truth, not to judge.

Of course, the most rational of you may say, "Isn't it strictly equivalent? Isn't telling students what they should have done the same thing as telling them that they were wrong?" Well, not quite. From a purely logical point of view, sure: if a question has only two possible answers, A or B, and the student wrongly chooses A, telling him that the correct answer is B is exactly the same as telling him, "You're wrong." And, by the same reasoning, in a binary fifty-fifty choice, strictly equivalent amounts of learning should occur upon hearing "You're right" and "You're wrong." Let's not forget, however, that children are not perfect logicians. For them, the additional step of deducing "If I chose A and I was wrong, then the correct answer must have been B" is not so immediate. On the other hand, they have no trouble grasping the main message: I messed up. Actually, when this very experiment was performed, adults succeeded in extracting equal amounts of information from reward and punishment, but adolescents did not: they learned much better from their successes than from their failures.[15] So, let us spare them this distress and give them the most neutral and informative feedback possible. Error feedback should not be confused with punishment.

GRADES, A POOR SUBSTITUTE FOR ERROR FEEDBACK

I must now say a few words about an educational institution that is full of defects, and yet so deeply rooted in tradition that we have a hard time imagining school without it: grades. According to learning theory, a grade is just a reward (or punishment!) signal. However, one of its obvious shortcomings is that it is totally lacking in precision. The grade of an exam is usually just a simple sum—and as such, it summarizes different sources of errors without distinguishing them. It is therefore insufficiently informative: by itself, it says nothing about the reason *why* we made a mistake, or *how* to correct ourselves. In the most extreme case, an F that stays an F provides zero information, only the clear social stigma of incompetence.

Grades alone, when not accompanied by detailed and constructive assessments, are therefore a poor source of error feedback. Not only are they imprecise, but they are also often delayed by several weeks, at which point most students have long forgotten which aspects of their inner reasoning misled them.

Grades can also be profoundly unfair, especially for students who are unable to keep up, because the level of the exams usually increases from week to week. Let's take the analogy of video games. When you discover a new game, you initially have no idea how to progress effectively. Above all, you don't want to be constantly reminded of how bad you are! That's why video game designers start with extremely easy levels, where you are almost sure to win. Very gradually, the difficulty increases and, with it, the risk of failure and frustration—but programmers know how to mitigate this by mixing the easy with the difficult, and by leaving you free to retry the same level as many times as you need. You see your score steadily increase . . . and finally, the joyous day comes when you successfully pass the final level, where you were stuck for so long. Now compare this with the report cards of "bad" students: they start the year off with a bad grade, and instead of motivating them by letting them take the same test again until they pass, the teacher gives them a new exercise every week, almost always beyond their abilities. Week after week, their "score" hovers around zero. In the video game market, such a design would be a complete disaster.

All too often, schools use grades as punishments. We cannot ignore the tremendous negative effects that bad grades have on the emotional systems of the brain: discouragement, stigmatization, feelings of helplessness. . . . Let us listen to the insightful voice of a professional dunce: Daniel Pennac, today a leading French writer who received the famous Renaudot Prize in 2007 for his book *School Blues*, but who was at the bottom of his class year after year:

> My school report cards confirmed this to me every month: if I was an idiot, it was only of my own making. Hence the self-hatred, the inferiority complex and, above all, the guilt. . . . I considered myself less than nothing. Because a good-for-nothing student, as my teachers repeatedly told me, *is* nothing. . . . I did not see any future for myself, and I had no possible representation of myself as an adult. Not because I didn't want anything, but because I thought that I was unfit for anything.[16]

Pennac eventually overcame this harmful state of mind (after flirting with suicide), but few children exhibit such resilience. The effects of school-induced stress have been particularly studied in the field of mathematics, the school subject most famous for the all-too-well-known anxiety it induces in so many students. In math class, some children suffer from a genuine form of math-induced depression because they know that, whatever they do, they will be punished with failure. Mathematics anxiety is a well-recognized, measured, and quantified syndrome. Children who suffer from it show activation in the pain and fear circuits, including the amygdala, which is located deep in the brain and is involved in negative emotions.[17] These students are not necessarily less intelligent than others, but the emotional tsunami that they experience destroys their abilities for calculation, short-term memory, and especially learning.

Numerous studies, both in humans and animals, confirm that stress and anxiety can dramatically hinder the ability to learn.[18] In the hippocampus of mice, for instance, fear conditioning literally solidifies neuronal plasticity: after the animal has been traumatized by random, unpredictable electrical shocks, the circuit finds itself in a similar state as toward the end of the sensitive period, when synapses have become immobile and frozen, entangled in rigid perineuronal nets. Conversely, being immersed in a fear-free, stimulating environment can reopen synaptic plasticity, thus freeing the neurons and returning their synaptic contacts to their childlike motility—a fountain of youth.

Giving out bad grades and presenting them as punishments, therefore, severely risks inhibiting children's progress, because stress and discouragement will prevent them from learning. In the long run, it can also alter their personality and self-image. The American psychologist Carol Dweck has largely studied the negative effects of this mental disposition, which consists of attributing one's failures (or successes) to a fixed, immutable aspect of one's personality—what she calls a "fixed mindset." "I'm bad at math," "Foreign languages are not my forte," and so on and so forth. She contrasts this view with the fundamentally correct idea that all children are capable of progress—what she calls a "growth mindset."

Her research suggests that, all other factors being equal, mindset plays an important role in learning.[19] Having a deeply entrenched view that anyone can progress is, in itself, a source of progress. Conversely, children who adhere to the idea that skills are immutable, and that one is either gifted or not, perform worse. Indeed, such a fixed mindset is demotivating: it encourages neither attention nor active engagement, and it interprets errors as markers of intrinsic inferiority. As we have seen, however, making errors is the most natural thing—it simply proves that we have tried. Remember Theodore Roosevelt: "The only man who never makes a mistake is the man who never does anything." Imagine if Grothendieck had come to the conclusion, at eleven years old, that he was bad at math because he thought that pi equalled three.

Research shows that even successful students can suffer from the fixed-mindset attitude. They too need to work in order to maintain their motivation, and we are not doing them any favors by letting them believe that because they are "gifted," they do not have to work hard.

Implementing a growth mindset does not mean telling every child that he or she is the best, under the simple pretext of nurturing their self-esteem. Rather, it means drawing attention to their day-to-day progress, encouraging their participation, rewarding their efforts . . . and, indeed, explaining to them the very foundations of learning: that all children have to make efforts, that they must always try out a response, and that erring (and correcting their errors) is the only way to learn.

Let's leave the last word to Daniel Pennac: "Teachers are not there to scare their students, but to help them overcome the fear of learning. Once this fear has been overcome, students' hunger for knowledge is insatiable."

TEST THYSELF

If grades are hardly effective, then what is the best way to incorporate our scientific knowledge of error processing into our classrooms? The rules are simple. First, students must be encouraged to participate, to put forth responses, to actively generate hypotheses, however tentative; and second, they must quickly receive objective, non-punitive feedback that allows them to correct themselves.

There is a strategy that meets all these criteria, and all teachers know about it: it is called . . . testing! What is less well-known is that dozens of scientific publications demonstrate its effectiveness. Regularly testing students' knowledge, a method referred to as "retrieval practice," is one of the most effective educational strategies.[20] Regular testing maximizes long-term learning. The mere act of putting your memory to the test makes it stronger. It is a direct reflection of the principles of active engagement and error feedback. Taking a test forces you to face reality head-on, to strengthen what you know, and to realize what you don't know.

The idea that testing is a cornerstone of the learning process is not self-evident. Most teachers and students see tests as a simple means of grading—their role is merely to assess the knowledge which has been acquired elsewhere, during class or while studying. Such ranking or grading, however, turns out to be the least interesting part of the test. What matters isn't the final grade that you get, but the effort that you make to retrieve information and the immediate feedback that you receive. In this respect, research shows that tests often play at least as important a role as the class itself.

This conclusion was attained in a famous series of experiments by the American psychologist Henry Roediger and his collaborators. In one study, they asked students to memorize words in a fixed amount of time, but with several different strategies. One group was told to spend all their time studying, in eight short sessions. A second group received six sessions of studying, interrupted by two tests. Finally, the third group alternated four brief study sessions and four tests. Because all three groups had the same amount of time, testing actually reduced the time available for studying. Yet the results were clear: forty-eight hours later, the students' memory of the word list was better the more opportunities they had to test themselves. Regularly alternating periods of studying and testing forced them to engage and receive explicit feedback ("I know this word now, but it's this other one I can never remember . . ."). Such self-awareness, or "meta-memory," is useful because it allows the learner to focus harder on the difficult items during the subsequent study sessions.[21] The effect is clear: the more you test yourself, the better you remember what you have to learn.

Here is another example: Imagine that you have to learn some words in a foreign language, such as *qamutiik*, the Inuit word for "sled." One possibility is to write the two words side by side on a card, in order to associate them mentally. Alternatively, you could read the Inuit word first and then, after five seconds, the translation. Note that the second condition reduces the amount of information available: during the first five seconds, you see only the word *qamutiik*, without being reminded what it means. However, it is this strategy that works best.[22] Why? Because it forces you to think first, to try to remember the meaning of the word before you receive feedback. Once again, active engagement followed by error feedback maximizes learning.

The paradox is that neither students nor their teachers are aware of these effects. If you ask their opinion, everyone thinks that testing oneself is a

distraction, and that studying is what matters. This is why students and teachers alike predict exactly the opposite of what is observed experimentally:
according to them, the more we study, the better we do. And in agreement
with this wrong idea, most students spontaneously spend their time reading
and rereading class notes and textbooks, highlighting each line with a different
color of the rainbow . . . all strategies that are much less effective than taking
a brief test.

Why do we have the illusion that cramming for an exam is the best
learning strategy? Because we are unable to differentiate between the various compartments of our memory. Immediately after reading our textbook or our class notes, information is fully present in our mind. It sits in
our conscious working memory, in an active form. We feel as if we know
it, because it is present in our short-term storage space . . . but this short-
term compartment has nothing to do with the long-term memory that we
will need in order to retrieve the same information a few days later. After
a few seconds or minutes, working memory already starts dissipating, and
after a few days, the effect becomes enormous: unless you retest your knowledge, memory vanishes. To get information into long-term memory, it is
essential to study the material, then test yourself, rather than spend all your
time studying.

It's easy to put these ideas into practice on your own. All you have to do
is prepare flash cards: on one side, you write a question, and on the other, the
answer. To test yourself, draw the cards one after the other, and for each card,
try to remember the answer (prediction) before checking it by turning to the
other side (error feedback). If you get the wrong answer, put the card back
toward the top of the pile—this will force you to revisit the same information
soon. If you get the right answer, put the card at the bottom of the pile: there
is no immediate need to study it again, but it will reappear sooner or later, at
a time when forgetting will have begun to take effect. There are now many
phone and tablet apps that allow you to build your own collection of flash
cards, and a similar algorithm underlies learning software, such as the famous
Duolingo for foreign languages.

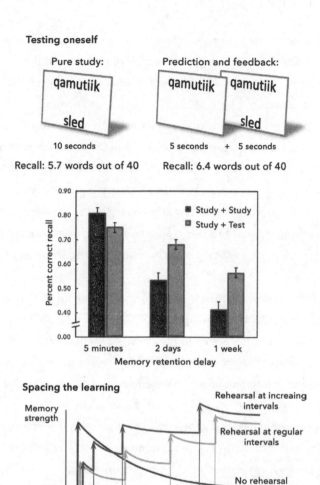

Self-testing is one of the best learning strategies, because it forces us to become aware of our mistakes. When learning foreign words, it is better to start off by trying to remember the word before receiving error feedback than to merely study each pair (top). Experiments also show that it is better to alternate periods of studying and testing than to spend all of one's time studying (middle). In the long run, memory is much better when rehearsal periods are spaced out, especially if the time intervals are gradually increased (bottom).

THE GOLDEN RULE: SPACING OUT THE LEARNING

Why does the alternation of studying and testing have such positive effects? Because it exploits one of the most effective strategies that educational science has discovered: the spacing out of training sessions. This is the golden rule: it is always better to spread out the training periods rather than cram them into a single run. The best way to ensure retention in the long term is with a series of study periods, interspersed with tests and spaced at increasingly large intervals.

Decades of psychological research show that if you have a fixed amount of time to learn something, spacing out the lessons is a much more effective strategy than grouping them.[23] The distribution of learning over several days has a tremendous effect: experiments show that you can multiply your memory by a factor of three when you review at regular intervals, rather than trying to learn everything at once. The rule is simple, and all musicians know it: fifteen minutes of work every day of the week is better than two hours on a single day per week.

Why is the spacing strategy so efficient? Brain imaging[24] shows that cramming the problems into a single session decreases the brain activity they evoke, perhaps because repeated information gradually loses its novelty. Repetition also seems to create an illusion of knowledge, an overconfidence due to the presence of information in working memory: it seems available, we have it in mind, so we do not see the point of working any harder. On the other hand, spacing out the learning increases brain activity: it seems to create an effect of "desirable difficulty" by prohibiting simple storage in working memory, and thus forcing the relevant circuits to work more.

What is the most effective time interval between two repetitions of the same lesson? A strong improvement is observed when the interval reaches twenty-four hours—probably because sleep, as we will see in a moment, plays a central role in consolidating what we learn. But American psychologist Hal Pashler and his colleagues have shown that the optimal interval depends on the desired duration of memory retention. If you need to remember the information for only a few days or weeks, then it is ideal to review it every day for about a week. If, on the other hand, knowledge must be maintained for several months or years, the revision interval should be extended proportionately. The rule of thumb is to review the information at intervals of approximately 20 percent of the desired memory duration—for instance, rehearse after two months if you want a memory to last about ten months. The effect is substantial: a single repetition of a lesson at a delay of a few weeks triples the number of items that can be recalled a few months later! To keep the information in memory as long as possible, it is best to gradually increase the time intervals themselves: start with rehearsals every day, then review the information after a week, a month, then a year. . . . This strategy guarantees optimal memory at all points in time.[25]

The figure above shows you why: each review reinforces learning. It refreshes the strength of mental representations and helps fight the exponential forgetfulness that characterizes our memory. Above all, the spacing out of learning sessions seems to select, out of all the available memory circuits in our brain, the one with the slowest forgetting curve, that is, the one that projects the information farthest into the future.

Indeed, we have been wrong about memory: it is not a system which is oriented toward the past, but one whose role is to send data to the future, so that we may later access it. By repeating the same information several times, at long intervals, we help our brain convince itself that this information is valuable enough to be delivered to our future self.

Hal Pashler draws several practical lessons from this research. First, learning always benefits from being spread over several sessions. Second, for school topics, reviewing after a few days or weeks is not enough. If you want to memorize something in the long run, you should review it after an interval of at least a few months. From this perspective, we have to rethink the entire organization of textbooks. Most of them are organized into chapters that focus on a specific topic (which is good) and are followed by questions and problems that focus only on that lesson (which is less good). This organization has two negative consequences: the lessons are not reviewed regularly or with sufficient spacing, and the exercises are dumbed-down, because students do not have to determine for themselves what knowledge or strategies should be used to address a given problem. Experiments show that it is better to mix all sorts of different problems, instead of limiting oneself to the most recent lesson, in order to regularly put all of one's knowledge to the test.[26]

What about finals or end-of-year exams? The science of learning suggests that they are not ideal, because they encourage last-minute work rather than regular practice. Nevertheless, they are still a useful test of acquired knowledge. Last-minute studying is not necessarily ineffective: provided the student has already made efforts to learn in the preceding months, intense study just before an exam refreshes the knowledge in memory and will help it last. However, a regular review of knowledge, year after year, is likely to yield even greater benefit. Short-term exams, which focus only on what was learned in the preceding weeks, do not guarantee long-term memory. A cumulative review, covering the entire program from the beginning of the year, works much better.

What is the point—you may be asking—of students studying the same things over the course of the school year? Why make them repeat an exercise that they have already completed several times? If they get perfect scores, will they learn anything at all? Of course they will. This may seem paradoxical in a chapter devoted to the benefits of error, but the benefit of feedback is not limited to items that students get wrong. On the contrary, receiving feedback improves memory even when the right answer was chosen.[27] Why? Because as long as knowledge is not perfectly consolidated, the brain continues to learn, even weakly. As long as there is uncertainty, error signals continue to spread in our brain. The difference between the initial low-confidence answer and the subsequent 100-percent-certain information acts as a useful feedback signal: it flags a virtual error that we could have made and from which we can therefore learn.

This is why overlearning is always beneficial: until our knowledge is absolutely certain, reviewing and testing it continues to improve our performance, especially in the long run. Moreover, repetition has other benefits for our brain: it automates our mental operations until they become unconscious. This is the last pillar of learning that remains to be examined: consolidation.

CHAPTER 10

Consolidation

CONSIDER A FIRST GRADER WHO SUCCESFULLY DEPLOYED THE THREE
pillars of learning and quickly learned to read. He actively engaged in reading, with curiosity and enthusiasm. He learned to pay attention to every letter
of every word, from left to right. And, over the months, as his errors receded,
he began to accurately decipher the correspondence between letters and
sounds and to store the spellings of irregular words. However, he is not a fluid
reader yet, and reads slowly and with effort. What's missing? He still has to
deploy the fourth pillar of learning: consolidation. His reading, which, at this
stage, mobilizes all his attention, has to become automatic and unconscious.

The analysis of his reading times is revealing: the longer a word is, the
longer it takes him to decipher it (see figure 18 in the color insert). The function is linear: response time increases by a fixed amount of about one-fifth of
a second for each additional letter. This is characteristic of a serial, step-by-
step operation—and it is completely normal: at his age, reading relies on deciphering letters or letter groups one by one, in a slow and attention-demanding
manner.[1] But this dysfluent phase should not last forever: with practice, in the
two years that follow, the child's reading will accelerate and become more
fluid. After two or three years of intensive practice, the effect of word length
will flat out disappear. Dear reader, at this very moment, as your expert brain

deciphers my words, you take the same exact amount of time to read any word between three and eight letters long. It takes, on average, about three years of training for visual word recognition to move from sequential to parallel. Ultimately, our visual word form area processes all the letters of a word simultaneously rather than serially.

This is an excellent example of the consolidation which happens in all domains: a shift from slow, conscious, and effortful processing to fast, unconscious, and automatic expertise. Our brains never stop learning. Even when a skill is mastered, we continue to overlearn it. Automatization mechanisms "compile" the operations we regularly use into more efficient routines. They transferred them to other brain circuits, outside our conscious awareness, where processes can unfold independent of one another without disrupting other operations in progress.

FREEING UP BRAIN RESOURCES

When you scan the brain of a beginner reader, what do you see? In addition to activation of the normal reading circuit—which includes visual areas for letter recognition and temporal-lobe areas for phoneme, syllable, and word processing—a massive activation of parietal and prefrontal regions is also present.[2] This intense and energy-hungry activity, reflecting effort, attention, and conscious executive control, will gradually disappear as learning consolidates (see figure 18 in the color insert). In an expert reader, these regions no longer contribute to reading—they are activated only if you disturb reading, for example by spacing out the l e t t e r s , or by ᴚɐ them, forcing the expert brain to revert to the slow, beginner mode.[3]

Automating reading means setting up a restricted and specialized circuit for the efficient processing of the strings of letters that we regularly encounter. As we learn, we develop an extraordinarily effective circuit for recognizing the most common characters as well as their combinations.[4] Our brain compiles statistics: it determines which letters are most frequent, where they appear most often, and in which associations they occur. Even the primary visual cortex adapts to the shapes and positions of the most frequent letters.[5] After a few years of overlearning, this circuit goes into routine mode and manages to function without the slightest conscious intervention.[6] At this stage, the activation of the parietal and prefrontal cortex has vanished: we can now read effortlessly.

What is true for reading also applies to all other areas of learning. Whether we learn to type, play a musical instrument, or drive a car, our gestures are initially under the control of the prefrontal cortex: we produce them slowly and consciously, one by one. Practice, however, makes perfect: over time, all effort evaporates, and we can exercise those skills while talking or thinking about something else. Repeated practice turns control over to the motor cortex and especially the basal ganglia, a set of subcortical circuits that record our automatic and routine behaviors (including prayers and swearing!). The same shift happens for arithmetic. For a beginner child, each calculation problem is an Everest that requires great effort to climb and mobilizes the circuits of the prefrontal cortex. At this stage, calculation is sequential: to solve 6 + 3, children will typically count the steps one by one: "Six . . . seven, eight . . . nine!" As consolidation progresses, children begin to retrieve the result straight from memory, and the prefrontal activity fades away in favor of specialized circuits in the parietal and ventral temporal cortex.[7]

Why is automatization so important? Because it frees up the cortex's resources. Remember that the parietal and prefrontal executive cortices operate as a generic executive control network that imposes a cognitive bottleneck: it cannot multitask. While our brain's central executive is focused on one task, all other conscious decisions are delayed or canceled. Thus, as long as a mental operation remains effortful, because it has not yet been automated by overlearning, it absorbs valuable executive attention resources and prevents us from focusing on anything else. Consolidation is essential because it makes our precious brain resources available for other purposes.

Let us take a concrete example. Imagine if you had to solve a math problem, but your reading had remained at the beginner's level: "A dryver leevz Bawstin att too oh clok and heds four Noo Yiorque too hunjred myels ahwey.

Hee ar eye-vz at ate oh clok. Wat waz hiz avrij speed?" I think you get my point: it is practically impossible to do both things at the same time. The difficulty of reading destroys any capacity for arithmetic reflection. To progress, it is essential that the mental tools most useful to us, such as reading or arithmetic, become second nature—that they operate unconsciously and effortlessly. We cannot reach the highest levels of the educational pyramid without first consolidating its foundations.

THE KEY ROLE OF SLEEP

We have already seen that learning is much more efficient when done at regular intervals: rather than cramming an entire lesson into one day, we are better off spreading out the learning. The reason is simple: every night, our brain consolidates what it has learned during the day. This is one of the most important neuroscience discoveries of the last thirty years: sleep is not just a period of inactivity or a garbage collection of the waste products that the brain accumulated while we were awake. Quite the contrary: while we sleep, our brain remains active; it runs a specific algorithm that replays the important events it recorded during the previous day and gradually transfers them into a more efficient compartment of our memory.

The discovery dates back to 1924. That year, two American psychologists, John Jenkins (1901–48) and Karl Dallenbach (1887–1971), revisited the classical studies on memory.[8] They re-examined the work of the pioneer of memory, German Hermann Ebbinghaus (1850–1909), who, as early as the end of the nineteenth century, had discovered a basic psychological law: the more time goes by, the less you remember what you learned. The Ebbinghaus forgetting curve is a beautiful, monotonously decreasing exponential. What Jenkins and Dallenbach noticed, however, is that the curve presented a single anomaly: it showed no memory loss between eight and fourteen hours after learning something new. Jenkins and Dallenbach had an epiphany: in the Ebbinghaus experiment, the eight-hour time limit corresponded to tests taken on the same day, and the fourteen-hour time limit to tests spaced one night apart.

To get to the bottom of this, they designed a new experiment that disentangled these two variables: the time elapsed before memory was tested, and whether or not the participants had the opportunity to sleep. To do so, they taught their students random syllables either around midnight, just before going to sleep, or in the morning. The result was clear: what we learn in the morning fades away with time, according to Ebbinghaus's exponential law; what is learned at midnight, on the other hand, remains stable over time (provided the students had at least two hours of sleep). In other words, sleeping prevents forgetting.

Several alternative interpretations of these results come to mind. Perhaps memory decays during the day because, while awake, the brain accumulates toxic substances that are eliminated during sleep; or perhaps memory suffers from interference with other events that occur in the interval between learning and testing, which does not happen during sleep. But these alternatives were definitively rejected in 1994, when Israeli researchers demonstrated that sleep *causes* additional learning: without any extra training, cognitive and motor performance improved after a period of sleep.[9] The experiment was simple. During the day, volunteers learned to detect a bar at a specific point on the retina. Their performance improved slowly, and it plateaued after a few hours of training: the limit seemed to have been reached. Send the participants to sleep, however, and surprise: when they wake up the next morning, their performance is much improved, and remains so throughout the following days. Sleep demonstrably causes the extra learning, because if we wake subjects up during the night each time they enter REM sleep, they show no improvement in the morning.

Numerous studies have confirmed and extended these early discoveries.[10] The amount of nightly gain varies according to the quality of sleep, which can be assessed by placing electrodes on the scalp and monitoring the slow waves that characterize deep sleep. Both the duration and the depth of sleep predict a person's performance improvement upon waking. The relationship also operates in the converse direction: the need for sleep seems to depend on the amount of stimulation and learning that occurred during the previous day. In

animals, a gene involved in cerebral plasticity, zif-268, increases its expression
in the hippocampus and cortex during REM sleep, specifically when the animals
were previously exposed to an enriched environment: the increased stimulation leads to a surge in nocturnal brain plasticity.[11]

The respective roles of the different stages of sleep are not yet perfectly
established, but it seems that deep sleep allows for the consolidation and generalization of knowledge (what psychologists call semantic or declarative
memory), while REM sleep, during which brain activity is close to a state of
wakefulness, reinforces perceptual and motor learning (procedural memory).

THE SLEEPING BRAIN RELIVES THE PREVIOUS DAY

While the psychological demonstrations of the effects of sleep were quite
convincing, the neural mechanism by which a sleeping brain could learn, even
better than while awake, remained to be identified. In 1994, neurophysiolo-
gists Matthew Wilson and Bruce McNaughton made a remarkable discovery:
in the absence of any external stimulation, neurons in the hippocampus spontaneously activate during sleep.[12] And this activity is not random: it retraces
the footsteps that the animal took during the day!

As we saw in Chapter 4, the hippocampus contains place cells, i.e., neurons that fire when an animal is (or believes itself to be) at a certain point in
space. The hippocampus is packed with a variety of place-coding neurons,
each of which prefers a different location. If you record enough of them, you
find that they span the entire space in which the animal walks. When a rat
moves through a corridor, some neurons fire at the entrance, others in the
middle, and yet others toward the end. Thus, the path that the rat takes is
reflected by the successive firing of a whole series of place cells: movement in
actual space becomes a temporal sequence in neural space.

And this is where Wilson and McNaughton's experiments fit in. They
discovered that when the rat falls asleep, the place cells in its hippocampus
start firing again, in the same order. The neurons literally replay the trajectories of the preceding wake period. The only difference is speed: during sleep,

neuronal discharges can be accelerated by a factor of twenty. In their sleep, rats dream of a high-speed race through their environment!

The relationship between the firing of hippocampal neurons and the position of the animal is so faithful that neuroscientists have managed to reverse the process, decoding the content of a dream from the animal's neuronal firing patterns.[13] During wakefulness, as the animal walks around in the real world, the systematic mapping between its location and its brain activity is recorded. These data make it possible to train a decoder, a computer program that reverses the relationship and guesses the animal's position from the pattern of neuronal firing. When this decoder is applied to sleep data, we see that while the animal dozes, its brain traces out virtual trajectories in space.

The rat's brain thus replays, at a high speed, the patterns of activity it experienced the day before. Every night brings back memories of the day. And such replay is not confined to the hippocampus, but extends to the cortex, where it plays a decisive role in synaptic plasticity and the consolidation of learning. Thanks to this nocturnal reactivation, even a single event of our lives, recorded only once in our episodic memory, can be replayed hundreds of times during the night (see figure 19 in the color insert). Such memory transfer may even be the main function of sleep.[14] It is possible that the hippocampus specializes in the storage of the events of the preceding day, using a fast single-trial learning rule. During the night, the reactivation of these neuronal signals spreads them to other neural networks, mainly located in the cortex and capable of extracting as much information as possible from each episode. Indeed, in the cortex of a rat that learns to perform a new task, the more a neuron reactivates during the night, the more it increases its participation in the task during the following day.[15] Hippocampal reactivation leads to cortical automation.

Does the same phenomenon exist in humans? Yes. Brain imaging shows that during sleep, the neural circuits that we used during the preceding day get reactivated.[16] After playing hours of Tetris, gamers were scanned the following night: they literally hallucinated a cascade of geometric shapes in their dreams, and their eyes made corresponding movements, from top to bottom.

What's more, in a recent study, volunteers fell asleep in an MRI machine and were suddenly awakened as soon as their electroencephalogram suggested that they were dreaming. The MRI showed that many areas of their brains had spontaneously activated just before they were woken, and that the recorded activity predicted the content of their dreams. If a participant reported, for instance, the presence of people in their dream, the experimenters detected sleep-induced activity in the cortical area associated with face recognition. Other experiments showed that the extent of this reactivation predicts not only the content of the dream, but also the amount of memory consolidation after waking up. Some neurosurgeons are even beginning to record single neurons in the human brain, and they see that, as in rats, their firing patterns trace out the sequence of events experienced on the preceding day.

Sleep and learning are strongly linked. Numerous experiments show that spontaneous variations in the depth of sleep correlate with variations in performance on the next day. When we learn to use a joystick, for example, during the following night, the frequency and intensity of slow sleep waves increase in the parietal regions of the brain involved in such sensorimotor learning—and the stronger the increase, the more a person's performance improves.[17] Similarly, after motor learning, brain imaging shows a surge of activity in the motor cortex, hippocampus, and cerebellum, accompanied by a decrease in certain frontal, parietal, and temporal areas.[18] Experiment after experiment gives convergent results: after sleeping, brain activity shifts around, and a portion of the knowledge acquired during the day is strengthened and transferred to more automatic and specialized circuits.

Although automation and sleep are tightly related, every scientist knows that correlation is not causation. Is the link a causal one? To verify this, we can artificially increase the depth of sleep by creating a resonance effect in the brain. During sleep, brain activity oscillates spontaneously at a slow frequency, on the order of forty to fifty cycles per minute. By giving the brain a small additional kick at just the right frequency, we can make these rhythms resonate and increase their intensity—a bit like when we push a swing at just the right moments, until it oscillates with a huge amplitude. German sleep scientist Jan Born did precisely this in two different ways: by passing tiny currents through the skull, and by simply playing a sound synchronized with the brain waves of the sleeper. Whether electrified or soothed by the sound of waves, the sleeping person's brain was carried away by this irresistible rhythm and produced significantly more slow waves characteristic of deep sleep. In both cases, on the following day, this resonance led to a stronger consolidation of learning.[19]

A French start-up has begun exploiting this effect: it sells headbands which supposedly facilitate sleep and increase the depth of sleep by playing quiet sounds that stimulate the slow rhythms of the nocturnal brain. Other researchers attempt to increase learning by forcing the brain to reactivate certain memories at night. Imagine learning certain facts in a classroom heavily scented with the smell of roses. Once you enter deep sleep, we spray your bedroom with the same fragrance. Experiments indicate that the information you learned is much better consolidated the next morning than if you had slept while being exposed to another smell.[20] The perfume of roses serves as an unconscious cue that biases your brain to reactivate this particular episode of the day, thus increasing its consolidation in memory.

The same effect can be achieved with auditory cues. Imagine that you are asked to memorize the locations of fifty images, each associated with a given sound (a cat meows, a cow moos, etc.). Fifty items are a lot to remember . . . but the night is there to help. In one experiment, during the night, the researchers stimulated the subjects' brains with half of the sounds. Hearing them unconsciously during deep sleep biased the nocturnal neuronal replay—and the next morning, the participants remembered the locations of the corresponding images much better.[21]

In the future, will we all fiddle with our sleep in order to learn better? Many students already do this spontaneously: they review an important lesson just before falling asleep, unknowingly attempting to bias their nocturnal replay. But let's not confuse such useful strategies with the misconception that one can acquire entirely new skills while sleeping. Some charlatans sell audio recordings that are supposed to teach you a foreign language unconsciously while you sleep. The research is clear—such tapes have no effect whatsoever.[22] Although there might be a few exceptions, the bulk of the evidence suggests that the sleeping brain does not absorb new information: it can only replay what it has already experienced. To learn a skill as complex as a new language, the only thing that works is practice during the day, then sleep during the night to reactivate and consolidate what we acquired.

DISCOVERIES DURING SLEEP

Does sleeping merely strengthen memory? Many scientists think otherwise: they report making discoveries during the night. The most famous case is the German chemist August Kekule von Stradonitz (1829–96), who first dreamed up the structure of benzene—an unusual molecule, because its six carbon atoms form a closed loop, like a ring or . . . a snake that bites its tail. This is how Kekule described his dream on that fateful night:

> Again the atoms were gamboling before my eyes. . . . My mental eye, rendered more acute by repeated visions of this kind, could now distinguish larger structures of manifold conformation; long rows sometimes more closely fitted together, all twining and twisting in snake-like motion. But look! What was that? One of the snakes had seized hold of its own tail, and the form whirled mockingly before my eyes.

And Kekule concluded: "Let us learn to dream, gentlemen, and then perhaps we shall learn the truth."

Can sleep really increase our creativity and lead us to truth? While science historians are divided on the authenticity of Kekule's Ouroboros episode, the idea of a nightly incubation is widespread among scientists and artists. The designer Philippe Starck said with humor in a recent interview, "Every night after putting my book down . . . I say to my wife: 'I'm off to work.'"[23] I myself have often had the experience of discovering the solution to a difficult problem upon waking up. However, a collection of anecdotes does not make for a proof. You have to experiment—and that's exactly what Jan Born and his team did.[24] During the day, these researchers taught volunteers a complex algorithm, which required applying a series of calculations to a given number. However, unbeknownst to the participants, the problem contained a hidden shortcut, a trick that cut the calculation time by a large amount. Before going to sleep, very few subjects had figured it out. However, a good night's sleep doubled the number of participants who discovered the shortcut, while those who were prevented from sleeping never experienced such a eureka moment. Moreover, the results were the same regardless of the time of day at which participants were tested. Thus, elapsed time was not the determining factor: only sleep led to genuine insight.

Nocturnal consolidation is therefore not limited to the strengthening of existing knowledge. The discoveries from the day are not only stored, but also recoded in a more abstract and general form. Nighttime neuronal replay undoubtedly has a crucial role in this process. Every night, our floating ideas from the day are reactivated hundreds of times at an accelerated rate, thus multiplying the chances that our cortex eventually discovers a rule that makes sense. In addition, the twentyfold acceleration of neural discharges compresses information. High-speed replay implies that the neurons that were activated at long intervals while awake now find themselves adjacent in the night sequence. This mechanism seems ideal for gathering, synthesizing, compressing, and "converting raw information into useful and exploitable knowledge"—the very definition of intelligence according to artificial intelligence mogul Demis Hassabis.

In the future, will intelligent machines have to sleep like we do? The question seems crazy, yet I think that, in a certain sense, they will: their learning algorithms will probably incorporate a consolidation phase similar to what we call sleep. Indeed, computer scientists have already designed several learning algorithms that mimic the sleep/wake cycle.[25] These algorithms provide inspiring models for the new vision of learning that I defend in this book, in which learning consists of building an internal generative model of the outside world. Remember that our brain contains massive internal models, capable of resynthesizing a variety of truer-than-life mental images, realistic dialogues, and meaningful deductions. In the awake state, we adjust these models to our environment: we use the sensory data that we receive from the outside world to select whichever model best fits the world around us. During this stage, learning is primarily a bottom-up operation: the unexpected incoming sensory signals, when confronted with the predictions of our internal models, generate prediction-error signals that climb up the cortical hierarchy and adjust the statistical weights at each step, so that our top-down models progressively gain in accuracy.

The new idea is that during sleep, our brain works in the opposite direction: from top to bottom. During the night, we use our generative models to synthesize new, unanticipated images, and part of our brain trains itself on this array of images created from scratch. This enhanced training set allows us to refine our ascending connections. Because both the parameters of the generative model and its sensory consequences are known, it is now much easier to discover the link between them. This is how we become more and more effective in extracting the abstract information that lies behind a specific sensory input: after a good night's sleep, the slightest clue suffices to identify the best mental model of reality, however abstract it may be.

According to this idea, dreams are nothing more than an enhanced training set of images: our brain relies on internal reconstructions of reality to multiply its necessarily limited experience of the day. Sleep seems to solve a problem that all learning algorithms face: the scarcity of the data available for training. To learn, current artificial neural networks need huge data sets— but life is too short, and our brain has to make do with the limited amount of information it can gather during the day. Sleep may be the solution that the brain found to simulate, in an accelerated manner, myriad events that an entire life would not suffice to experience for real.

During these thought experiments, we occasionally make discoveries. There is nothing magical about this: as our mental simulation engine runs, it sometimes hits upon unexpected outcomes—a bit like a chess player, once she has mastered the rules, can spend years exploring their consequences. Indeed, humanity owes to mental imagery some of its greatest scientific discoveries— when Einstein dreamed of riding a photon, for instance, or when Newton imagined the moon falling onto the earth like an apple. Even Galileo's most famous experiment, in which he dropped objects from the Tower of Pisa to prove that their free-falling speed does not depend on their mass, probably never took place. A thought experiment sufficed: Galileo imagined dropping two spheres, one light and one heavy, from the top of the tower; supposed that the heavier one would fall faster; and used his mental models to show that this led to a contradiction. Suppose, he said, that I connect the two spheres with a wire of negligible mass. The resulting two-sphere system, now forming a heavier object, should fall even faster. But this is absurd because the lighter sphere, which falls less quickly, should slow down the heavier one. These never-ending contradictions lead to only one possibility: all objects fall at the same speed regardless of their mass.

This is the kind of reasoning that our mental simulator affords, day or night. The very fact that we can conjure such complex mental scenes highlights the extraordinary array of algorithms in our brain. Of course, we learn during the day, but nocturnal neuronal replay multiplies our potential. This may indeed be one of the secrets of the human species, because suggestive data indicate that our sleep may be the deepest and most effective of all primates.[26]

SLEEP, CHILDHOOD, AND SCHOOL

What about children? Everyone knows that infants spend most of their time sleeping, and that sleep shortens with age. This is logical: early childhood is a privileged period during which our learning algorithms have a heavier workload. In fact, experimental data show that, for the same length of time, a child's sleep is two to three times more effective than that of an adult. After intensive learning, ten-year-old children dive much faster into deep sleep than adults. Their slow waves are more intense, and the result is clear: when they study a sequence, sink into sleep, and wake up the next day refreshed and rested, they discover more regularities than adults.[27]

Nocturnal consolidation is already at work during the first few months of life. Infants under one year of age rely on it, for example, when they learn a novel word. Babies who take a short nap, only an hour and a half long, retain much better the words that they learned within the few hours before falling asleep.[28] Above all, they generalize them better: the first time babies hear the word "horse," they associate it only with one or two specific instances of horses, but after having slept, their brains manage to associate this word with new specimens that they have never seen before. Like Kekule in the crib, these budding scientists make discoveries during their sleep and wake up with a much better theory of the word *horse*.

What about school-age children? Research is equally clear: in preschool, even a brief afternoon nap strengthens the memory of what the children learned in the morning.[29] For maximum benefit, sleep should occur within hours of learning. This benefit, however, exists only in children who regularly take naps. Since the brain naturally regulates its need for sleep according to the stimulation of the day, it does not seem useful to force children to nap, but we should encourage napping for those who feel the need.

Unfortunately, with TV, smartphones, and internet galore, children's sleep, like that of adults, is now threatened on all fronts. What are the consequences? Can chronic sleep deprivation go so far as to cause specific learning disabilities, which are apparently on the rise? This is still only a hypothesis, but there are some suggestive hints.[30] For instance, a subset of hyperactive children with attention disorders may simply be suffering from a chronic lack of sleep. Some experience sleep apneas that prevent them from falling into deep sleep—and simply clearing out the airways suffices to eliminate not only their chronic sleep deficit, but also their attention impairment. Recent experiments even suggest that electrical stimulation of the brain, by increasing the depth of slow sleep waves, may mitigate the learning deficit in hyperactive children.

Let me be clear: these recent data still need to be replicated, and I am in no way denying the existence of genuine attention disorders (in children for whom attention training, or sometimes the drug Ritalin, can have very positive effects). From an educational perspective, however, there is little doubt that improving the length and quality of sleep can be an effective intervention for all children, especially those with learning difficulties.

This idea has been tested in teenagers. Around puberty, chronobiology shows that the sleep cycle shifts: adolescents do not feel the need to go to bed early, but, as everyone may have experienced, they have the greatest difficulty getting up. It is not that they are unwilling so much as a simple consequence of the massive turmoil in the neural and hormonal networks that control their sleep/wake cycle. Unfortunately, no one seems to have informed school principals, who continue to require students to be present early in the morning. What would be so bad about changing this arbitrary convention? The experiment has been done, with promising results: once the start of school is delayed by half an hour to an hour, teenagers get more sleep, school attendance increases, attention in class improves, and grades shoot up.[31] And the list of positive effects could go on: the American Academy of Pediatrics strongly recommends delaying school start times as an efficient countermeasure to teenage obesity, depression, and accidents (e.g., drowsy driving). That children's general physical and mental well-being can be so easily improved, at strictly no cost, provides a magnificent example of adapting the educational system to the constraints of brain biology.

CONCLUSION

Reconciling Education
with Neuroscience

The greatest and most important difficulty of human science
is the nurture and education of children.

Montaigne, *Essays* (1580)

Pedagogy is like medicine: an art, but one which is based—or
should be based—on precise scientific knowledge.

Jean Piaget, "La pédagogie moderne" (1949)

AT THE END OF THIS JOURNEY, I HOPE TO HAVE CONVINCED YOU THAT,
thanks to recent advances in cognitive psychology, neuroscience, artificial
intelligence, and education sciences, we now possess detailed knowledge
about how our brain learns. This knowledge is not self-evident, and most of
our preconceived ideas about learning need to be rescinded:

- No, babies are not blank slates: as early as the first year of life, they possess
  vast knowledge of objects, numbers, probabilities, space, and people.
- No, the child's brain is not a sponge that obediently absorbs the structure
  of its environment. Remember Felipe, the blind and tetraplegic Brazilian

storyteller, or Nicholas Saunderson, the blind mathematician who held Newton's chair: such cases show us that sensory inputs can be disrupted or absent without ruining a child's grasp of abstract ideas.

- No, the brain is not just a network of malleable neurons that waits to be shaped by its inputs: all the large fiber bundles are present at birth, and brain plasticity, however indispensable, typically refines only the last millimeters of our connections.

- No, learning does not occur passively through simple exposure to data or lectures: on the contrary, cognitive psychology and brain imaging show us that children are budding scientists, constantly generating new hypotheses, and that the brain is an ever-alert organ that learns by testing the models it projects onto the outside world.

- No, errors are not the mark of bad students: making mistakes is an integral part of learning, because our brain can adjust its models only when it discovers a discrepancy between what it envisioned and reality.

- No, sleep is not just a period of rest: it is an integral part of our learning algorithm, a privileged period during which our brain plays its models in a loop and enhances the experience of the day by a factor of ten to one hundred.

- And no, today's learning machines are nowhere close to surpassing the human brain: our brains remain, for the moment at least, the fastest, most effective, and most energy efficient of all information processing devices. A true probabilistic machine, it successfully extracts the maximum amount of information from each moment of the day and transforms it at night into abstract and general knowledge, in a way that we do not yet know how to reproduce in computers.

In the Promethean battle between the computer chip and the neuron, the machine and the brain, the latter still has the advantage. For sure, in principle, there is nothing in the mechanics of the brain that a machine could not imitate. Indeed, all the ideas I have exposed here are already in the hands of computer scientists whose research is overtly inspired by neuroscience.[1] In practice, however, machines still have a long way to go. To improve, they will need many of the ingredients that we reviewed here: an internal language of thought that allows concepts to be flexibly recombined; algorithms that reason with probability distributions; a curiosity function; effective systems for managing attention and memory; and perhaps a sleep/wake algorithm that expands the training set and increases the chances of discovery. Algorithms of this type are beginning to appear, but they remain light years away from the performance of a newborn baby. The brain keeps the upper hand over machines, and I predict that it will for a long time.

THIRTEEN TAKE-HOME MESSAGES TO OPTIMIZE CHILDREN'S POTENTIAL

The more I study the human brain, the more I am impressed. But I also know that its performance is fragile, as it strongly depends on the environment in which it develops. Too many children do not reach their full potential because their families or schools do not provide them with ideal conditions for learning.

International comparisons are alarming: they show that, over the past fifteen or twenty years, the school results of many Western countries, including my home country, France, have plunged, while those of many Asian countries and cities—such as Singapore, Shanghai, and Hong Kong—have soared.[2] In mathematics, which used to be France's greatest strength, scores fell so sharply between 2003 and 2015 that my country now occupies the last place in Europe in the TIMSS survey, which evaluates the achievements of fifteen-year-old students in math and science.

Faced with such poor results, we are sometimes too quick to point our fingers at teachers. In reality, nobody knows the reasons behind this recent downfall: Are the culprits the parents, the schools, or society as a whole? Should we blame lack of sleep, inattention, or video games? Whatever the reasons may be, I am convinced that recent advances in the science of learning may help reverse this dark trend. We now know a lot more about the conditions that maximize learning and memory. All of us, parents and teachers alike, must learn to implement these conditions in our daily lives, at home and in the classroom.

The scientific results that I have presented converge toward simple, easily applicable ideas. Let's review them together:

- **Do not underestimate children.** At birth, infants possess a rich set of core skills and knowledge. Object concepts, number sense, a knack for languages, knowledge of people and their intentions . . . so many brain modules are already present in young children, and these foundational skills will later be recycled in physics, mathematics, language, and philosophy classes. Let us take advantage of children's early intuitions: each word and symbol that they learn, however abstract, must connect to prior knowledge. This connection is what will give them meaning.

- **Take advantage of the brain's sensitive periods.** In the first years of life, billions of synapses are created and destroyed every day. This effervescent activity makes the child's brain particularly receptive, especially for language learning. We should expose children to a second language as early as possible. We should also bear in mind that plasticity extends at least until adolescence. During this entire period, foreign language immersion can transform the brain.

- **Enrich the environment.** Learning wise, the child's brain is the most powerful of supercomputers. We should respect it by providing it with the right data at an early age: word or construction games, stories, puzzles. . . . Let's not hesitate to hold serious talks with our children, to answer their questions, even the most difficult, using an elaborate vocabulary, and to explain to them what we understand of the world. By giving our little ones an enriched environment, particularly regarding languages, we maximize their brain growth and prolong their juvenile plasticity.

- **Rescind the idea that all children are different.** The idea that each of us has a distinct learning style is a myth. Brain imaging shows that we all rely on very similar brain circuits and learning rules. The brain circuits for reading and mathematics are the same in each of us, give or take a few millimeters—even in blind children. We all face similar hurdles in learning, and the same teaching methods can surmount them. Individual differences, when they exist, lie more in children's extant knowledge, motivation, and the rate at which they learn. Let's carefully determine each child's current level in order to select the most relevant problems—but above all, let's ensure that all children acquire the fundamentals of language, literacy, and mathematics that everyone needs.

• **Pay attention to attention.** Attention is the gateway to learning: virtually no information will be memorized if it has not previously been amplified by attention and awareness. Teachers should become masters at capturing their students' attention and directing it to what matters. This implies carefully getting rid of any source of distraction: overly illustrated textbooks and excessively decorated classrooms only distract children from their primary task and prevent them from concentrating.

• **Keep children active, curious, engaged, and autonomous.** Passive students do not learn much. Make them more active. Engage their intelligence so that their minds sparkle with curiosity and constantly generate new hypotheses. But do not expect them to discover everything on their own: guide them through a structured curriculum.

• **Make every school day enjoyable.** Reward circuits are essential modulators of brain plasticity. Activate them by rewarding every effort and making every hour of class fun. No child is insensitive to material rewards—but their social brains respond equally to smiles and encouragement. The feeling of being appreciated and the awareness of one's own progress are rewards in and of themselves. Conversely, do away with the anxiety and stress that prevent learning—especially in mathematics.

• **Encourage efforts.** A pleasurable school experience is not synonymous with "effortless." On the contrary, the most interesting things to learn— reading, math, or playing an instrument—require years of practice. The belief that everything comes easy can lead children to think that they are dunces if they do not succeed. Explain to them that all students must try hard and that, when they do, everyone makes progress. Adopt a growth mindset, not a fixed mindset.

- **Help students deepen their thinking.** The deeper our brain processes information, the better we can remember. Never be content with superficial learning; always aim for deeper understanding. And remember Henry Roediger's words: "Making learning conditions more difficult, thus requiring students to engage more cognitive effort, often leads to enhanced retention."

- **Set clear learning objectives.** Students learn best when the purpose of learning is clearly stated to them and when they can see that everything at their disposal converges toward that purpose. Clearly explain what is expected of them, and stay focused on that goal.

- **Accept and correct mistakes.** To update their mental models, our brain areas must exchange error messages. Error is therefore the very condition of learning. Let us not punish errors, but correct them quickly, by giving children detailed but stress-free feedback. According to the Education Endowment Foundation's synthesis, the quality of the feedback that teachers provide to their students is the most effective lever for academic progress.

- **Practice regularly.** One-shot learning is not enough—children need to consolidate what they have learned to render it automatic, unconscious, and reflexive. Such routinization frees up our prefrontal and parietal circuits, allowing them to attend to other activities. The most effective strategy is to space out learning: a little bit every day. Spacing out practice or study sessions allows information to be permanently imprinted to memory.

- **Let students sleep.** Sleep is an essential ingredient of our learning algorithm. Our brain benefits each time we sleep, even when we nap. So, let us make sure that our children sleep long and deep. To get the most out of our brain's unconscious night work, studying a lesson or rereading a problem just before falling asleep can be a nifty trick. And because adolescents' sleep cycle is shifted, let's not wake them up too early!

Only by getting to know ourselves better can we make the most of the powerful algorithms with which our brains are equipped. All children would probably benefit from knowing the four pillars of learning: attention, active engagement, error feedback, and consolidation. Four slogans effectively summarize them: "Fully concentrate," "participate in class," "learn from your mistakes," and "practice every day, take advantage of every night." These are very simple messages that we should all heed.

AN ALLIANCE FOR THE SCHOOLS OF TOMORROW

How can we harmonize our school system with the discoveries of cognitive and brain sciences? A new alliance is needed. Just like medicine relies on a whole pyramid of biological and drug-design research, I believe that in the future, education will increasingly rely on evidence-based research, including fundamental laboratory experiments, as well as classroom-scale trials and deployment studies. Only by combining the distinct forces of teachers, parents, and scientists will we attain the worthy goal of reviving the curiosity and joy of learning in all children, in order to help them optimize their cognitive potential.

Experts of the classroom, teachers are entrusted with the priceless task of educating our children, who will soon have the future of this world in their hands. Yet we often leave teachers with very minimal resources to accomplish this goal. They deserve much greater respect and investment. Teachers today face increasingly severe challenges, including diminishing resources, expanding class sizes, growing violence, and the relentless tyranny of the curriculum. Amazingly, most teachers receive little or no professional training in the science of learning. My feeling is that we should urgently change this state of affairs, because we now possess considerable scientific knowledge about the brain's learning algorithms and the pedagogies that are the most efficient. I hope that this book can provide a small step toward a global revision of teacher training programs, in order to offer them the best tools from cognitive science, in line with their commitment to our children.

I hope that teachers will also agree that their pedagogical freedom should in no way be restricted by the growing science of the learning brain. On the contrary, one goal of this book is to allow them to better exercise this freedom. "I think of a hero," said Bob Dylan, "as someone who understands the degree of responsibility that comes with his freedom." Genuine pedagogical creativity can only come from full awareness of the range of available strategies and the ability to choose carefully from them, with full knowledge of their impact on students. The principles I have articulated throughout this book are compatible with multiple pedagogical approaches, and much can be done to put them into practice in the classroom. I expect a lot from teachers' inventiveness, because I think it is essential to children's enthusiasm.

In my opinion, the schools of the future should also have a much more important place for parents. They are the primary actors in a child's development, whose actions precede and prolong school. Home is where children have a chance to expand, through work and games, the knowledge that they acquired in class. Family is open seven days a week and, thus, can, better than school, take full advantage of each alternation of wakefulness and sleep, of learning and consolidation. Schools should devote more time to parent training, because this is one of the most effective interventions: well-trained parents can be invaluable teammates for teachers and astute observers of their children's difficulties.

Finally, scientists must engage with teachers and schools in order to consolidate the growing field of education science. Compared with the huge progress of the past thirty years in cognitive and brain sciences, educational research remains a relatively neglected area of study. Research organizations should encourage scientists to conduct major research programs in all areas of learning sciences, from neuroscience and brain imaging to the neuropsychology of developmental disorders, cognitive psychology, and educational sociology. Scaling up from the laboratory to the classroom is not as easy as it sounds, and we are in great need of full-scale experiments in schools. Cognitive science can help design and evaluate innovative educational tools.

Just as medicine is based on biology, the field of education must be grounded in a systematic and rigorous research ecosystem that brings together teachers, patients, and researchers, in a ceaseless search for more effective, evidence-based learning strategies.

ACKNOWLEDGMENTS

Many encounters stimulated the growth of this book. Twenty-five years ago, Michael Posner and Bruce McCandliss, then at the University of Oregon, were the first to convince me that cognitive science could be relevant to education. I owe much to the many scientific meetings they organized with the help of Bruno della Chiesa and the Organization for Economic Cooperation and Development (OECD). In the following decade, a wonderful group of South American friends—Marcela Peña, Sidarta Ribeiro, Mariano Sigman, Alejandro Maiche, and Juan Valle Lisboa—took the lead and trained an entire generation of young scientists at the unforgettable annual meetings of the Latin American School for Education, Cognitive and Neural Sciences. I am eternally grateful to them, as well as to the James S. McDonnell Foundation and its leaders, John Bruer and Susan Fitzpatrick, for giving me the chance to participate in all of them.

Another person who shared those stimulating experiences is my wife and colleague, Ghislaine Dehaene-Lambertz. We have been discussing brain development and, incidentally, the education of our children for thirty-two years. It goes without saying that I owe everything to her, including her meticulous reading of the preceding pages.

Another anniversary has gone by: it has been thirty-four years since I joined the laboratories of Jacques Mehler and Jean-Pierre Changeux. Their influence on my thinking is immense, and they will recognize many of their favorite themes in this book—as will other very close colleagues and friends, such as Lucia Braga, Laurent Cohen, Naama Friedmann, Véronique Izard, Régine Kolinsky, José Morais, Lionel Naccache, Christophe Pallier, Mariano Sigman, Elizabeth Spelke, and Josh Tenenbaum.

Thanks also go to my dear friend Antonio Battro, who continually encouraged me to pursue research on mind, brain, and education. I am also grateful to him for introducing me to Nico, an artist with a remarkable personality who very kindly allowed me to reproduce some of his paintings here. Thanks also to Yoshua Bengio, Alain Chédotal, Guillaume and David Dehaene, Molly Dillon, Jessica Dubois, György Gergely, Eric Knudsen, Leah Krubitzer, Bruce McCandliss, Josh Tenenbaum, Fei Xu, and Robert Zatorre for allowing me to reproduce the many figures in this book.

I would also like to thank all the institutions that have supported my research over the years with unfailing loyalty, in particular the Institut National de la Santé et de la Recherche Médicale (INSERM), Commissariat à l'Énergie Atomique et aux Énergies Alternatives (CEA), Collège de France, Université Paris-Sud, the European Research Council (ERC), and the Bettencourt Schueller Foundation. Thanks to them, I have been able to surround myself with brilliant and energetic students and collaborators. They are too numerous to list here, but they will recognize themselves in the long list of publications that follows. A special mention goes out to Anna Wilson, Dror Dotan, and Cassandra Potier-Watkins, with whom I developed educational software and classroom interventions.

Jean-Michel Blanquer, the French Ministre de l'Education Nationale, honored me with his trust by proposing that I chair his first Scientific Council, an exciting challenge for which I thank him wholeheartedly. I am grateful to all the members of the council, including Esther Duflo, Michel Fayol, Marc Gurgand, Caroline Huron, Elena Pasquinelli, Franck Ramus, Elizabeth Spelke, and Jo Ziegler, and my secretary-general Nelson Vallejo-Gomez, for their commitment and for all they have taught me.

This edition benefited greatly from the critical eye of my editors at Viking: Wendy Wolf and Terezia Cicel. And it wouldn't have made it into their hands without the ceaseless help of my agents, John and Max at Brockman Inc. Thank you for your constant support and invaluable feedback.

Yallingup, Australia, April 7, 2019

NOTES

INTRODUCTION

1. See the movies *The Miracle Worker* (1962) and *Marie's Story* (2014), as well as read the following books: Arnould, 1900; Keller, 1903.
2. Learning in the nematode *C. elegans*: Bessa, Maciel, and Rodrigues, 2013; Kano et al., 2008; Rankin, 2004.
3. Website of the Education Endowment Foundation (EEF): educationendowment foundation.org.uk.
4. The brain constantly keeps track of uncertainty: Meyniel and Dehaene, 2017; Heilbron and Meyniel, 2019.

CHAPTER 1: SEVEN DEFINITIONS OF LEARNING

1. You can try this experiment for yourself at the C3RV34U exhibition I organized at the Cité des sciences, Paris's main science museum.
2. LeNet artificial neural network: LeCun, Bottou, Bengio, and Haffner, 1998.
3. Visualizing the hierarchy of hidden units in the GoogLeNet artificial neural network: Olah, Mordvintsev, and Schubert, 2017.
4. Progressive separation of the ten digits by a deep neural network: Guerguiev, Lillicrap, and Richards, 2017.
5. Reinforcement learning: Mnih et al., 2015; Sutton and Barto, 1998.
6. Artificial neural network that learns to play Atari video games: Mnih et al., 2015.
7. Artificial neural network that learns to play Go: Banino et al., 2018; Silver et al., 2016.

8. Adversarial learning: Goodfellow et al., 2014.
9. Convolutional neural networks: LeCun, Bengio, and Hinton, 2015; LeCun et al., 1998.
10. Darwin's natural selection algorithm: Dennett, 1996.

CHAPTER 2: WHY OUR BRAIN LEARNS BETTER THAN CURRENT MACHINES

1. Artificial neural networks primarily implement the unconscious operations of the brain: Dehaene, Lau, and Kouider, 2017.
2. Artificial neural networks tend to learn superficial regularities: Jo and Bengio, 2017.
3. Generation of images that confuse humans as well as artificial neural networks: Elsayed et al., 2018.
4. Artificial neural network that learns to recognize CAPTCHAs: George et al., 2017.
5. Critique of the learning speed in artificial neural networks: Lake, Ullman, Tenenbaum, and Gershman, 2017.
6. Lack of systematicity in artificial neural networks: Fodor and Pylyshyn, 1988; Fodor and McLaughlin, 1990.
7. Language of thought hypothesis: Amalric, Wang, et al., 2017; Fodor, 1975.
8. Learning to count as program inference: Piantadosi, Tenenbaum, and Goodman, 2012; see also Piantadosi, Tenenbaum, and Goodman, 2016.
9. Recursive representations as a singularity of the human species: Dehaene, Meyniel, Wacongne, Wang, and Pallier, 2015; Everaert, Huybregts, Chomsky, Berwick, and Bolhuis, 2015; Hauser, Chomsky, and Fitch, 2002; Hauser and Watumull, 2017.
10. Human singularity in coding an elementary sequence of sounds: Wang, Uhrig, Jarraya, and Dehaene, 2015.
11. Acquisition of geometrical rules—slow in monkeys, ultrafast in children: Jiang et al., 2018.
12. The conscious human brain resembles a serial Turing machine: Sackur and Dehaene, 2009; Zylberberg, Dehaene, Roelfsema, and Sigman, 2011.
13. Fast learning of word meaning: Tenenbaum, Kemp, Griffiths, and Goodman, 2011; Xu and Tenenbaum, 2007.
14. Word learning based on shared attention: Baldwin et al., 1996.
15. Knowledge of determiners and other function words at twelve months: Cyr and Shi, 2013; Shi and Lepage, 2008.
16. Mutual exclusivity principle in word learning: Carey and Bartlett, 1978; Clark, 1988; Markman and Wachtel, 1988; Markman, Wasow, and Hansen, 2003.
17. Reduced reliance on mutual exclusivity in bilinguals: Byers-Heinlein and Werker, 2009.
18. Rico, a dog who learned hundreds of words: Kaminski, Call, and Fischer, 2004.
19. Modelling of an "artificial scientist": Kemp and Tenenbaum, 2008.

20. Discovering the causality principle: Goodman, Ullman, and Tenenbaum, 2011; Tenenbaum et al., 2011.
21. The brain as a generative model: Lake, Salakhutdinov, and Tenenbaum, 2015; Lake et al., 2017.
22. Probability theory is the logic of science: Jaynes, 2003.
23. Bayesian model of information processing in the cortex: Friston, 2005. For empirical data on hierarchical passing of probabilistic error messages in the cortex, see, for instance, Chao, Takaura, Wang, Fujii, and Dehaene, 2018; Wacongne et al., 2011.

CHAPTER 3: BABIES' INVISIBLE KNOWLEDGE

1. Object concept in infants: Baillargeon and DeVos, 1991; Kellman and Spelke, 1983.
2. Fast acquisition of how objects fall, and what suffices to keep them supported: Baillargeon, Needham, and DeVos, 1992; Hespos and Baillargeon, 2008.
3. Number concept in infants: Izard, Dehaene-Lambertz, and Dehaene, 2008; Izard, Sann, Spelke, and Streri, 2009; Starkey and Cooper, 1980; Starkey, Spelke, and Gelman, 1990. A detailed review of these findings can be found in the second edition of my book *The Number Sense* (Dehaene, 2011).
4. Multimodal knowledge of numbers in neonates: Izard et al., 2009.
5. Small-number addition and subtraction in infants: Koechlin, Dehaene, and Mehler, 1997; Wynn, 1992.
6. Large-number addition and subtraction in infants: McCrink and Wynn, 2004.
7. The accuracy of number sense gets refined with age and education: Halberda and Feigenson, 2008; Piazza et al., 2010; Piazza, Pica, Izard, Spelke, and Dehaene, 2013.
8. Number sense in chicks: Rugani, Fontanari, Simoni, Regolin, and Vallortigara, 2009; Rugani, Vallortigara, Priftis, and Regolin, 2015.
9. Number neurons in untrained animals: Ditz and Nieder, 2015; Viswanathan and Nieder, 2013.
10. Brain-imaging and single-cell evidence for number neurons in humans: Piazza, Izard, Pinel, Le Bihan, and Dehaene, 2004; Kutter, Bostroem, Elger, Mormann, and Nieder, 2018.
11. Core knowledge in infants: Spelke, 2003.
12. Bayesian reasoning in infants: Xu and Garcia, 2008.
13. The child as a "scientist in the crib": Gopnik, Meltzoff, and Kuhl, 1999; Gopnik et al., 2004.
14. Infants' understanding of probabilities, containers, and randomness: Denison and Xu, 2010; Gweon, Tenenbaum, and Schulz, 2010; Kushnir, Xu, and Wellman, 2010.
15. Babies distinguish whether a machine or a human draws from a container: Ma and Xu, 2013.
16. Logical reasoning in twelve-month-old babies: Cesana-Arlotti et al., 2018.

17. Infants' understanding of intentions: Gergely, Bekkering, and Király, 2002; Gergely and Csibra, 2003; see also Warneken and Tomasello, 2006.

18. Ten-month-old infants infer other people's preferences: Liu, Ullman, Tenenbaum, and Spelke, 2017.

19. Babies evaluate other people's actions: Buon et al., 2014.

20. Babies distinguish intentional and accidental actions: Behne, Carpenter, Call, and Tomasello, 2005.

21. Face processing by fetuses in utero: Reid et al., 2017.

22. Face recognition in infancy and development of cortical responses to faces: Adibpour, Dubois, and Dehaene-Lambertz, 2018; Deen et al., 2017; Livingstone et al., 2017.

23. Face recognition in the first year of life: Morton and Johnson, 1991.

24. Babies prefer to listen to their maternal language: Mehler et al., 1988.

25. "The baby in my womb leaped for joy": Luke 1:44.

26. See my book *Consciousness and the Brain* (2014).

27. Lateralization of language and voice processing in premature babies: Mahmoudzadeh et al., 2013.

28. Word segmentation in infants: Hay, Pelucchi, Graf Estes, and Saffran, 2011; Saffran, Aslin, and Newport, 1996.

29. Young children detect grammatical violations: Bernal, Dehaene-Lambertz, Millotte, and Christophe, 2010.

30. Limits of language-learning experiments in animals: see, for instance, Penn, Holyoak, and Povinelli, 2008; Terrace, Petitto, Sanders, and Bever, 1979; Yang, 2013.

31. Fast emergence of language in deaf communities: Senghas, Kita, and Özyürek, 2004.

CHAPTER 4: THE BIRTH OF A BRAIN

1. Brain imaging of language in infants: Dehaene-Lambertz et al., 2006; Dehaene-Lambertz, Dehaene, and Hertz-Pannier, 2002.

2. Empiricist view of the infant's brain: see, for instance, Elman et al., 1996; Quartz and Sejnowski, 1997.

3. Evolution of cortical areas (figure 7 in the color insert): Krubitzer, 2007.

4. Hierarchy of cortical responses to language in humans: Lerner, Honey, Silbert, and Hasson, 2011; Pallier, Devauchelle, and Dehaene, 2011.

5. Organization of major long-range cortical fiber tracts at birth: Dehaene-Lambertz and Spelke, 2015; Dubois et al., 2015.

6. Hypothesis of a disorganized brain that receives the imprint of the environment: Quartz and Sejnowski, 1997.

7. The peripheral nervous system is already remarkably organized by two months of gestation: Belle et al., 2017.

8. Subdivision of the cortex into Brodmann areas: Amunts et al., 2010; Amunts and Zilles, 2015; Brodmann, 1909.
9. Early gene expression in delimited cortical areas: Kwan et al., 2012; Sun et al., 2005.
10. Early origins of brain asymmetries: Dubois et al., 2009; Leroy et al., 2015.
11. Brain asymmetries in left- and right-handers: Sun et al., 2012.
12. Self-organizing model of cortical folds: Lefevre and Mangin, 2010.
13. Grid cells in rats: Banino et al., 2018; Brun et al., 2008; Fyhn, Molden, Witter, Moser, and Moser, 2004; Hafting, Fyhn, Molden, Moser, and Moser, 2005.
14. Self-organizing models of grid cells: Kropff and Treves, 2008; Shipston-Sharman, Solanka, and Nolan, 2016; Widloski and Fiete, 2014; Yoon et al., 2013.
15. Fast emergence of grid cells, place cells, and head direction cells during development: Langston et al., 2010; Wills, Cacucci, Burgess, and O'Keefe, 2010.
16. Grid cells in humans: Doeller, Barry, and Burgess, 2010; Nau, Navarro Schröder, Bellmund, and Doeller, 2018.
17. Spatial navigation in a blind child: Landau, Gleitman, and Spelke, 1981.
18. Fast emergence of cortical areas for faces versus places: Deen et al., 2017; Livingstone et al., 2017.
19. Tuning to numbers in parietal cortex: Nieder and Dehaene, 2009.
20. Self-organizing model of number neurons: Hannagan, Nieder, Viswanathan, and Dehaene, 2017.
21. Self-organization based on an internal "game engine in the head": Lake et al., 2017.
22. Genes and cell migration in dyslexia: Galaburda, LoTurco, Ramus, Fitch, and Rosen, 2006.
23. Connectivity anomalies in dyslexia: Darki, Peyrard-Janvid, Matsson, Kere, and Klingberg, 2012; Hoeft et al., 2011; Niogi and McCandliss, 2006.
24. Phonological predictors of dyslexia in six-month-old children: Leppanen et al., 2002; Lyytinen et al., 2004.
25. Attentional dyslexia: Friedmann, Kerbel, and Shvimer, 2010.
26. Visual dyslexia with mirror errors: McCloskey and Rapp, 2000.
27. Bell curve for dyslexia: Shaywitz, Escobar, Shaywitz, Fletcher, and Makuch, 1992.
28. Cognitive and neurological impairments in dyscalculia: Butterworth, 2010; Iuculano, 2016.
29. Parietal gray-matter loss in premature children with dyscalculia: Isaacs, Edmonds, Lucas, and Gadian, 2001.

CHAPTER 5: NURTURE'S SHARE

1. Synaptic hypothesis of brain plasticity: Holtmaat and Caroni, 2016; Takeuchi, Duszkiewicz, and Morris, 2014.
2. Music activates reward circuits: Salimpoor et al., 2013.

3. Long-term potentiation of synapses: Bliss and Lømo, 1973; Lømo, 2018.
4. Aplysia, hippocampus, and synaptic plasticity: Pittenger and Kandel, 2003.
5. Hippocampus and memory for places: Whitlock, Heynen, Shuler, and Bear, 2006.
6. Memory for fearful sounds in mice: Kim and Cho, 2017.
7. Causal role of synaptic changes: Takeuchi et al., 2014.
8. Nature of the engram, the neuronal basis of a memory: Josselyn, Köhler, and Frankland, 2015; Poo et al., 2016.
9. Working memory and sustained firing: Courtney, Ungerleider, Keil, and Haxby, 1997; Ester, Sprague, and Serences, 2015; Goldman-Rakic, 1995; Kerkoerle, Self, and Roelfsema, 2017; Vogel and Machizawa, 2004.
10. Working memory and fast synaptic changes: Mongillo, Barak, and Tsodyks, 2008.
11. Role of the hippocampus in the fast acquisition of novel information: Genzel et al., 2017; Lisman et al., 2017; Schapiro, Turk-Browne, Norman, and Botvinick, 2016; Shohamy and Turk-Browne, 2013.
12. Displacement of a memory engram from hippocampus to cortex: Kitamura et al., 2017.
13. Creation of a false memory in mice: Ramirez et al., 2013.
14. Turning a bad memory into a good one: Ramirez et al., 2015.
15. Erasing a traumatic memory: Kim and Cho, 2017.
16. Creating a novel memory during sleep: de Lavilléon et al., 2015.
17. Tool and symbol learning in macaque monkeys: Iriki, 2005; Obayashi et al., 2001; Srihasam, Mandeville, Morocz, Sullivan, and Livingstone, 2012.
18. Distant synaptic changes: Fitzsimonds, Song, and Poo, 1997.
19. Anatomical changes due to music training: Gaser and Schlaug, 2003; Oechslin, Gschwind, and James, 2018; Schlaug, Jancke, Huang, Staiger, and Steinmetz, 1995.
20. Anatomical changes due to literacy: Carreiras et al., 2009; Thiebaut de Schotten, Cohen, Amemiya, Braga, and Dehaene, 2014.
21. Anatomical changes after learning to juggle: Draganski et al., 2004; Gerber et al., 2014.
22. Brain changes in London taxi drivers: Maguire et al., 2000, 2003.
23. Non-synaptic mechanism of memory in the cerebellum: Johansson, Jirenhed, Rasmussen, Zucca, and Hesslow, 2014; Rasmussen, Jirenhed, and Hesslow, 2008.
24. Effects of physical exercise and nutrition on the brain: Prado and Dewey, 2014; Voss, Vivar, Kramer, and van Praag, 2013.
25. Cognitive deficits in children with vitamin B1 (thiamine) deficiency: Fattal, Friedmann, and Fattal-Valevski, 2011.
26. Brain plasticity in a child born without a right hemisphere: Muckli, Naumer, and Singer, 2009.
27. Turning auditory cortex into visual cortex: Sur, Garraghty, and Roe, 1988; Sur and Rubenstein, 2005.

28. Hypothesis of a disorganized brain that receives the imprint of the environment: Quartz and Sejnowski, 1997.
29. Self-organization of visual maps by retinal waves: Goodman and Shatz, 1993; Shatz, 1996.
30. Progressive adjustment of cortical spontaneous activity: Berkes, Orbán, Lengyel, and Fiser, 2011; Orbán, Berkes, Fiser, and Lengyel, 2016.
31. Review of the concept of sensitive periods: Werker and Hensch, 2014.
32. Growth of human cortical neurons: Conel, 1939; Courchesne et al., 2007.
33. Synaptic overproduction and elimination in the course of development: Rakic, Bourgeois, Eckenhoff, Zecevic, and Goldman-Rakic, 1986.
34. Distinct phases of synaptic elimination in humans: Huttenlocher and Dabholkar, 1997.
35. Progressive myelination of cortical bundles: Dubois et al., 2007, 2015; Flechsig, 1876.
36. Acceleration of visual responses in babies: Adibpour et al., 2018; Dehaene-Lambertz and Spelke, 2015.
37. Slowness of conscious processing in babies: Kouider et al., 2013.
38. Sensitive period for binocular vision: Epelbaum, Milleret, Buisseret, and Duffer, 1993; Fawcett, Wang, and Birch, 2005; Hensch, 2005.
39. Loss of the capacity to discriminate non-native phonemes: Dehaene-Lambertz and Spelke, 2015; Maye, Werker, and Gerken, 2002; Pena, Werker, and Dehaene-Lambertz, 2012; Werker and Tees, 1984.
40. Partial recovery of the discrimination of /R/ and /L/ in Japanese speakers: McCandliss, Fiez, Protopapas, Conway, and McClelland, 2002.
41. Auditory cortex anatomy predicts the capacity to learn foreign contrasts: Golestani, Molko, Dehaene, Le Bihan, and Pallier, 2007.
42. Sensitive period for second-language learning: Flege, Munro, and MacKay, 1995; Hartshorne, Tenenbaum, and Pinker, 2018; Johnson and Newport, 1989; Weber-Fox and Neville, 1996.
43. Sharp decline in the speed of second-language grammar learning around seventeen years of age (analysis of data from several million people): Hartshorne et al., 2018.
44. Sensitive period for language learning in deaf people with a cochlear implant: Friedmann and Rusou, 2015.
45. Biological mechanisms for the opening and closing of sensitive periods: Caroni, Donato, and Muller, 2012; Friedmann and Rusou, 2015; Werker and Hensch, 2014.
46. Restoring brain plasticity: Krause et al., 2017.
47. Reorganization of language areas in adopted children: Pallier et al., 2003. Similar results have been observed in the domain of face recognition: when adopted in a Western country before the age of nine, Korean children lose the advantage that is

usually observed for recognizing members of one's own race (Sangrigoli, Pallier, Argenti, Ventureyra, and de Schonen, 2005).

48. Dormant trace of the first language in adopted children: Pierce, Klein, Chen, Delcenserie, and Genesee, 2014.
49. Dormant connections in owls: Knudsen and Knudsen, 1990; Knudsen, Zheng, and DeBello, 2000.
50. Age-of-acquisition effect in word processing: Ellis and Lambon Ralph, 2000; Gerhand and Barry, 1999; Morrison and Ellis, 1995.
51. Bucharest Early Intervention Project: Almas et al., 2012; Berens and Nelson, 2015; Nelson et al., 2007; Sheridan, Fox, Zeanah, McLaughlin, and Nelson, 2012; Windsor, Moraru, Nelson, Fox, and Zeanah, 2013.
52. Ethics of the Bucharest project: Millum and Emanuel, 2007.

CHAPTER 6: RECYCLE YOUR BRAIN

1. Nabokov, 1962.
2. Difficulties of illiterates in picture recognition: Kolinsky et al., 2011; Kolinsky, Morais, Content, and Cary, 1987; Szwed, Ventura, Querido, Cohen, and Dehaene, 2012.
3. Difficulties of illiterates in processing mirror images: Kolinsky et al., 2011, 1987; Pegado, Nakamura, et al., 2014.
4. Difficulties of illiterates in attending to part of a face: Ventura et al., 2013.
5. Difficulties of illiterates in recognizing and remembering spoken words: Castro-Caldas, Petersson, Reis, Stone-Elander, and Ingvar, 1998; Morais, 2017; Morais, Bertelson, Cary, and Alegria, 1986; Morais and Kolinsky, 2005.
6. Impact of arithmetic education: Dehaene, Izard, Pica, and Spelke, 2006; Dehaene, Izard, Spelke, and Pica, 2008; Piazza et al., 2013; Pica, Lemer, Izard, and Dehaene, 2004.
7. Counting and arithmetic in Amazon Indians: Pirahã: Frank, Everett, Fedorenko, and Gibson, 2008; Munduruku: Pica et al., 2004; Tsimane: Piantadosi, Jara-Ettinger, and Gibson, 2014.
8. Acquisition of the number line concept: Dehaene, 2003; Dehaene et al., 2008; Siegler and Opfer, 2003.
9. Neuronal recycling hypothesis: Dehaene, 2005, 2014; Dehaene and Cohen, 2007.
10. Evolution by duplication of brain circuits: Chakraborty and Jarvis, 2015; Fukuchi-Shimogori and Grove, 2001.
11. Learning confined to a neuronal subspace: Galgali and Mante, 2018; Golub et al., 2018; Sadtler et al., 2014.
12. One-dimensional coding in parietal cortex: Chafee, 2013; Fitzgerald et al., 2013.
13. Role of parietal cortex in the comparison of social status: Chiao, 2010.

14. Two-dimensional coding in entorhinal cortex: Yoon et al., 2013.

15. Coding of an arbitrary two-dimensional space by grid cells: Constantinescu, O'Reilly, and Behrens, 2016.

16. Coding of syntactic trees in Broca's area: Musso et al., 2003; Nelson et al., 2017; Pallier et al., 2011.

17. The number sense: Dehaene, 2011.

18. Number neurons in untrained animals: Ditz and Nieder, 2015; Viswanathan and Nieder, 2013.

19. Effect of training on number neurons: Viswanathan and Nieder, 2015.

20. Acquisition of Arabic numerals in monkeys: Diester and Nieder, 2007.

21. Relation between addition, subtraction, and movements of spatial attention: Knops, Thirion, Hubbard, Michel, and Dehaene, 2009; Knops, Viarouge, and Dehaene, 2009.

22. Functional MRI of professional mathematicians: Amalric and Dehaene, 2016, 2017.

23. Brain imaging of number processing in babies: Izard et al., 2008.

24. Functional MRI of early math in preschoolers: Cantlon, Brannon, Carter, and Pelphrey, 2006. Cantlon and Li, 2013, show that cortical areas for language and number are already active when a four-year-old watches the corresponding sections of *Sesame Street* movies, and that their activity predicts the child's language and math skills.

25. Blind mathematicians: Amalric, Denghien, and Dehaene, 2017.

26. Recycling of occipital cortex for math in the blind: Amalric, Denghien, et al., 2017; Kanjlia, Lane, Feigenson, and Bedny, 2016.

27. Language processing in the occipital cortex of the blind: Amedi, Raz, Pianka, Malach, and Zohary, 2003; Bedny, Pascual-Leone, Dodell-Feder, Fedorenko, and Saxe, 2011; Lane, Kanjlia, Omaki, and Bedny, 2015; Sabbah et al., 2016.

28. Debate on cortical plasticity in the blind: Bedny, 2017; Hannagan, Amedi, Cohen, Dehaene-Lambertz, and Dehaene, 2015.

29. Retinotopic maps in the blind: Bock et al., 2015.

30. Recycling of visual cortex in the blind: Abboud, Maidenbaum, Dehaene, and Amedi, 2015; Amedi et al., 2003; Bedny et al., 2011; Mahon, Anzellotti, Schwarzbach, Zampini, and Caramazza, 2009; Reich, Szwed, Cohen, and Amedi, 2011; Striem-Amit and Amedi, 2014; Strnad, Peelen, Bedny, and Caramazza, 2013.

31. Connectivity predicts function in visual cortex: Bouhali et al., 2014; Hannagan et al., 2015; Saygin et al., 2012, 2013, 2016.

32. Distance effect in number comparison: Dehaene, 2007; Dehaene, Dupoux, and Mehler, 1990; Moyer and Landauer, 1967.

33. Distance effect when deciding that two numbers are different: Dehaene and Akhavein, 1995; Diester and Nieder, 2010.

34. Distance effect when verifying addition and subtraction problems: Groen and Parkman, 1972; Pinheiro-Chagas, Dotan, Piazza, and Dehaene, 2017.

35. Mental representation of prices: Dehaene and Marques, 2002; Marques and Dehaene, 2004.
36. Mental representation of parity: Dehaene, Bossini, and Giraux, 1993; negative numbers: Blair, Rosenberg-Lee, Tsang, Schwartz, and Menon, 2012; Fischer, 2003; Gullick and Wolford, 2013; fractions: Jacob and Nieder, 2009; Siegler, Thompson, and Schneider, 2011.
37. Language of thought in mathematics: Amalric, Wang, et al., 2017; Piantadosi et al., 2012, 2016.
38. See my previous book *Reading in the Brain*: Dehaene, 2009.
39. Brain mechanisms of the invariant recognition of written words: Dehaene et al., 2001, 2004.
40. Connections between the visual word form area and language areas: Bouhali et al., 2014; Saygin et al., 2016.
41. Imaging of the illiterate brain: Dehaene et al., 2010; Dehaene, Cohen, Morais, and Kolinsky, 2015; Pegado, Comerlato, et al., 2014.
42. Specialization of early visual cortex for reading: Chang et al., 2015; Dehaene et al., 2010; Szwed, Qiao, Jobert, Dehaene, and Cohen, 2014.
43. Literacy competes with face processing the left hemisphere: Dehaene et al., 2010; Pegado, Comerlato, et al., 2014.
44. Development of reading and face recognition: Dehaene-Lambertz, Monzalvo, and Dehaene, 2018; Dundas, Plaut, and Behrmann, 2013; Li et al., 2013; Monzalvo, Fluss, Billard, Dehaene, and Dehaene-Lambertz, 2012.
45. Insufficient activity evoked by words and faces in dyslexic children: Monzalvo et al., 2012.
46. Universal marker of reading difficulties: Rueckl et al., 2015.
47. Competition between words and faces—knockout or blocking?: Dehaene-Lambertz et al., 2018.
48. Learning to read in adulthood: Braga et al., 2017; Cohen, Dehaene, McCormick, Durant, and Zanker, 2016.
49. Displacement of the visual word form area in musicians: Mongelli et al., 2017.
50. Reduced response to faces in mathematicians: Amalric and Dehaene, 2016.
51. Numerous long-term effects of early education: see the Abecedarian program (Campbell et al., 2012, 2014; Martin, Ramey, and Ramey, 1990), the Perry preschool program (Heckman, Moon, Pinto, Savelyev, and Yavitz, 2010; Schweinhart, 1993), and the Jamaican Study (Gertler et al., 2014; Grantham-McGregor, Powell, Walker, and Himes, 1991; Walker, Chang, Powell, and Grantham-McGregor, 2005).
52. Child-directed speech and vocabulary growth: Shneidman, Arroyo, Levine, and Goldin-Meadow, 2013; Shneidman and Goldin-Meadow, 2012.
53. Increased response to speech following parent-child story reading: Hutton et al., 2015, 2017; see also Romeo et al., 2018.

54. Advantages of early bilingualism: Bialystok, Craik, Green, and Gollan, 2009; Costa and Sebastián-Gallés, 2014; Li, Legault, and Litcofsky, 2014.

55. Benefits of an enriched environment: Donato, Rompani, and Caroni, 2013; Knudsen et al., 2000; van Praag, Kempermann, and Gage, 2000; Voss et al., 2013; Zhu et al., 2014.

CHAPTER 7: ATTENTION

1. Attention in mice: Wang and Krauzlis, 2018.
2. Attention in artificial neural networks: Bahdanau, Cho, and Bengio, 2014; Cho, Courville, and Bengio, 2015.
3. Attention in an artificial neural network learning to caption pictures (figure on page 149): Xu et al., 2015.
4. Inattention strongly reduces learning: Ahissar and Hochstein, 1993.
5. Reduced learning in the absence of attention and consciousness: Seitz, Lefebvre, Watanabe, and Jolicoeur, 2005; Watanabe, Nanez, and Sasaki, 2001.
6. Prefrontal ignition and access to consciousness: Dehaene and Changeux, 2011; van Vugt et al., 2018.
7. Acetylcholine, dopamine, brain plasticity, and alteration of cortical maps: Bao, Chan, and Merzenich, 2001; Froemke, Merzenich, and Schreiner, 2007; Kilgard and Merzenich, 1998.
8. Balance between inhibition and excitation, and reopening of brain plasticity: Werker and Hensch, 2014.
9. Activation of reward and alerting circuits by video games: Koepp et al., 1998.
10. Positive effects of video game training: Bavelier et al., 2011; Cardoso-Leite and Bavelier, 2014; Green and Bavelier, 2003.
11. Cognitive training using video games: see our math software at www.thenumberrace.com and www.thenumbercatcher.com; for reading acquisition, visit grapholearn.fr.
12. Spatial attention orienting: Posner, 1994.
13. Amplification by attention: Çukur, Nishimoto, Huth, and Gallant, 2013; Desimone and Duncan, 1995; Kastner and Ungerleider, 2000.
14. Inattentional blindness: Mack and Rock, 1998; Simons and Chabris, 1999.
15. Attentional blink: Marois and Ivanoff, 2005; Sergent, Baillet, and Dehaene, 2005.
16. Unattended items induce little or no learning: Leong, Radulescu, Daniel, DeWoskin, and Niv, 2017.
17. Adult experiment on attention to letters versus whole words: Yoncheva, Blau, Maurer, and McCandliss, 2010.
18. Educational studies of phonics versus whole-word reading: Castles, Rastle, and Nation, 2018; Ehri, Nunes, Stahl, and Willows, 2001; National Institute of Child Health and Human Development, 2000; see also Dehaene, 2009.

19. Organization of executive control in prefrontal cortex: D'Esposito and Grossman, 1996; Koechlin, Ody, and Kouneiher, 2003; Rouault and Koechlin, 2018.
20. Prefrontal expansion in the human species: Elston, 2003; Sakai et al., 2011; Schoenemann, Sheehan, and Glotzer, 2005; Smaers, Gómez-Robles, Parks, and Sherwood, 2017.
21. Prefrontal hierarchy and metacognitive control: Fleming, Weil, Nagy, Dolan, and Rees, 2010; Koechlin et al., 2003; Rouault and Koechlin, 2018.
22. Global neuronal workspace: Dehaene and Changeux, 2011; Dehaene, Changeux, Naccache, Sackur, and Sergent, 2006; Dehaene, Kerszberg, and Changeux, 1998; Dehaene and Naccache, 2001.
23. Central bottleneck: Chun and Marois, 2002; Marti, King, and Dehaene, 2015; Marti, Sigman, and Dehaene, 2012; Sigman and Dehaene, 2008.
24. Unawareness of the dual-task delay: Corallo, Sackur, Dehaene, and Sigman, 2008; Marti et al., 2012.
25. Debate on the ability to split attention and execute two tasks in parallel: Tombu and Jolicoeur, 2004.
26. An exceedingly decorated classroom distracts pupils: Fisher, Godwin, and Seltman, 2014.
27. Use of electronic devices in class reduces exam performance: Glass and Kang, 2018.
28. A-not-B error and development of prefrontal cortex: Diamond and Doar, 1989; Diamond and Goldman-Rakic, 1989.
29. Development of executive control and number perception: Borst, Poirel, Pineau, Cassotti, and Houdé, 2013; Piazza, De Feo, Panzeri, and Dehaene, 2018; Poirel et al., 2012.
30. Effect of number training on prefrontal cortex: Viswanathan and Nieder, 2015.
31. Role of executive control in cognitive and emotional development: Houdé et al., 2000; Isingrini, Perrotin, and Souchay, 2008; Posner and Rothbart, 1998; Sheese, Rothbart, Posner, White, and Fraundorf, 2008; Siegler, 1989.
32. Effects of training on executive control and working memory: Diamond and Lee, 2011; Habibi, Damasio, Ilari, Elliott Sachs, and Damasio, 2018; Jaeggi, Buschkuehl, Jonides, and Shah, 2011; Klingberg, 2010; Moreno et al., 2011; Olesen, Westerberg, and Klingberg, 2004; Rueda, Rothbart, McCandliss, Saccomanno, and Posner, 2005.
33. Randomized studies of Montessori pedagogy: Lillard and Else-Quest, 2006; Marshall, 2017.
34. Effects of musical training on the brain: Bermudez, Lerch, Evans, and Zatorre, 2009; James et al., 2014; Moreno et al., 2011.
35. Relation between executive control, prefrontal cortex, and intelligence: Duncan, 2003, 2010, 2013.
36. Training effects on fluid intelligence: Au et al., 2015.
37. Impact of adoption on IQ: Duyme, Dumaret, and Tomkiewicz, 1999.

38. Impact of education on IQ: Ritchie and Tucker-Drob, 2018.
39. Effects of cognitive training on concentration, reading, and arithmetic: Bergman-Nutley and Klingberg, 2014; Blair and Raver, 2014; Klingberg, 2010; Spencer-Smith and Klingberg, 2015.
40. Correlation between working memory and subsequent math scores: Dumontheil and Klingberg, 2011; Gathercole, Pickering, Knight, and Stegmann, 2004; Geary, 2011.
41. Joint training of working memory and the number line: Nemmi et al., 2016.
42. Learning Chinese with a nanny, but not with a video: Kuhl, Tsao, and Liu, 2003.
43. Shared attention and the pedagogical stance: Csibra and Gergely, 2009; Egyed, Király, and Gergely, 2013.
44. Object pointing and memory of object's identity: Yoon, Johnson, and Csibra, 2008.
45. Pseudo-teaching in meerkats: Thornton and McAuliffe, 2006.
46. Intelligent versus slavish copying of actions by fourteen-month-olds: Gergely et al., 2002.
47. Social conformism in perception: see, for instance, Bond and Smith, 1996.

CHAPTER 8: ACTIVE ENGAGEMENT

1. Classic experiment comparing active and passive kittens: Held and Hein, 1963.
2. Statistical learning of syllables and words: Hay et al., 2011; Saffran et al., 1996; see also ongoing research in G. Dehaene-Lambertz's lab on learning in sleeping neonates.
3. Effect of word processing depth on explicit memory: Craik and Tulving, 1975; Jacoby and Dallas, 1981.
4. Memory for sentences: Auble and Franks, 1978; Auble, Franks, and Soraci, 1979.
5. "Making learning conditions more difficult . . .": Zaromb, Karpicke, and Roediger, 2010.
6. Brain imaging of the effect of word processing depth on memory: Kapur et al., 1994.
7. The activation of prefrontal-hippocampal loops during incidental learning predicts subsequent memory: Brewer, Zhao, Desmond, Glover, and Gabrieli, 1998; Paller, McCarthy, and Wood, 1988; Sederberg et al., 2006; Sederberg, Kahana, Howard, Donner, and Madsen, 2003; Wagner et al., 1998.
8. Memory for conscious and unconscious words: Dehaene et al., 2001.
9. Active learning of physics concepts: Kontra, Goldin-Meadow, and Beilock, 2012; Kontra, Lyons, Fischer, and Beilock, 2015.
10. Comparison of traditional lecturing versus active learning: Freeman et al., 2014.
11. Failure of discovery learning and related pedagogical strategies: Hattie, 2017; Kirschner, Sweller, and Clark, 2006; Kirschner and van Merriënboer, 2013; Mayer, 2004.
12. To add all numbers from 1 to 100, pair 1 with 100, 2 with 99, 3 with 98, and so forth. Each of these pairs adds up to 101, and there are fifty of them, hence the total is 5050.

13. Instructional guidance rather than pure discovery: Mayer, 2004.
14. Urban legends in education: Kirschner and van Merriënboer, 2013.
15. The myth of learning styles: Pashler, McDaniel, Rohrer, and Bjork, 2008.
16. Variations in amount of reading in first grade: Anderson, Wilson, and Fielding, 1988.
17. Early childhood curiosity and academic achievement: Shah, Weeks, Richards, and Kaciroti, 2018.
18. Dopaminergic neurons sensitive to new information: Bromberg-Martin and Hikosaka, 2009.
19. Novelty seeking in rats: Bevins, 2001.
20. Brain imaging of curiosity: Gruber, Gelman, and Ranganath, 2014; see also Kang et al., 2009.
21. Laughter as an epistemic emotion unique to humans: Hurley, Dennett, and Adams, 2011.
22. Laughter and learning: Esseily, Rat-Fischer, Somogyi, O'Regan, and Fagard, 2016.
23. Review of psychological theories of curiosity: Loewenstein, 1994.
24. Inverted-U curve of curiosity: Kang et al., 2009; Kidd, Piantadosi, and Aslin, 2012, 2014; Loewenstein, 1994.
25. Curiosity in a robot: Gottlieb, Oudeyer, Lopes, and Baranes, 2013; Kaplan and Oudeyer, 2007.
26. Goldilocks effect in eight-month-olds: Kidd et al., 2012, 2014.
27. Metacognition in young children: Dehaene et al., 2017; Goupil, Romand-Monnier, and Kouider, 2016; Lyons and Ghetti, 2011.
28. Gender and race stereotypes in mathematics: Spencer, Steele, and Quinn, 1999; Steele and Aronson, 1995.
29. Stress, anxiety, learned helplessness, and the inability to learn: Caroni et al., 2012; Donato et al., 2013; Kim and Diamond, 2002; Noble, Norman, and Farah, 2005.
30. Explicit teaching may kill curiosity: Bonawitz et al., 2011.

CHAPTER 9: ERROR FEEDBACK

1. Grothendieck, 1986.
2. John Hattie's meta-analysis grants feedback an effect size of 0.73 standard deviations, which makes it one of the most powerful modulators of learning (Hattie, 2008).
3. Rescorla-Wagner learning rule: Rescorla and Wagner, 1972.
4. For a detailed criticism of associative learning, see Balsam and Gallistel, 2009; Gallistel, 1990.
5. Blocking of animal conditioning: Beckers, Miller, De Houwer, and Urushihara, 2006; Fanselow, 1998; Waelti, Dickinson, and Schultz, 2001.
6. Surprise enhances infants' learning and exploration: Stahl and Feigenson, 2015.

7. Error signals in the brain: Friston, 2005; Naatanen, Paavilainen, Rinne, and Alho, 2007; Schultz, Dayan, and Montague, 1997.

8. Surprise reflects the violation of a prediction: Strauss et al., 2015; Todorovic and de Lange, 2012.

9. Hierarchy of local and global error signals: Bekinschtein et al., 2009; Strauss et al., 2015; Uhrig, Dehaene, and Jarraya, 2014; Wang et al., 2015.

10. Surprise due to an unexpected picture: Meyer and Olson, 2011.

11. Surprise due to a semantic violation: Curran, Tucker, Kutas, and Posner, 1993; Kutas and Federmeier, 2011; Kutas and Hillyard, 1980.

12. Surprise due to a grammatical violation: Friederici, 2002; Hahne and Friederici, 1999; but see also Steinhauer and Drury, 2012, for a critical discussion.

13. Prediction error in the dopamine network: Pessiglione, Seymour, Flandin, Dolan, and Frith, 2006; Schultz et al., 1997; Waelti et al., 2001.

14. Importance of high-quality feedback at school: Hattie, 2008.

15. Learning by trial and error in adults versus adolescents: Palminteri, Kilford, Coricelli, and Blakemore, 2016.

16. Pennac, D. (2017, February 11). Daniel Pennac: "J'ai été d'abord et avant tout professeur." *Le Monde*. Retrieved from lemonde.fr.

17. Math anxiety syndrome: Ashcraft, 2002; Lyons and Beilock, 2012; Maloney and Beilock, 2012; Young, Wu, and Menon, 2012.

18. Effect of fear conditioning on synaptic plasticity: Caroni et al., 2012; Donato et al., 2013.

19. Fixed versus growth mindset: Claro, Paunesku, and Dweck, 2016; Dweck, 2006; Rattan, Savani, Chugh, and Dweck, 2015. Note, however, that the size of these effects, and therefore their practical relevance at school, has been recently questioned: Sisk, Burgoyne, Sun, Butler, and Macnamara, 2018.

20. Massive effect of retrieval practice on learning: Carrier and Pashler, 1992; Karpicke and Roediger, 2008; Roediger and Karpicke, 2006; Szpunar, Khan, and Schacter, 2013; Zaromb and Roediger, 2010. For an excellent review of the relative efficacy of various learning techniques, see Dunlosky, Rawson, Marsh, Nathan, and Willingham, 2013.

21. Making retrospective memory judgments facilitates learning: Robey, Dougherty, and Buttaccio, 2017.

22. Retrieval practice facilitates the acquisition of foreign vocabulary: Carrier and Pashler, 1992; Lindsey, Shroyer, Pashler, and Mozer, 2014.

23. Spacing the learning improves memory retention: Cepeda et al., 2009; Cepeda, Pashler, Vul, Wixted, and Rohrer, 2006; Rohrer and Taylor, 2006; Schmidt and Bjork, 1992.

24. Brain imaging of the spacing effect: Bradley et al., 2015; Callan and Schweighofer, 2010.

25. Effect of progressively increasing the time between lessons: Kang, Lindsey, Mozer, and Pashler, 2014.

26. The shuffling of mathematics problems improves learning: Rohrer and Taylor, 2006, 2007.

27. Feedback improves memory even on correct trials: Butler, Karpicke, and Roediger, 2008.

CHAPTER 10: CONSOLIDATION

1. Moving from serial to parallel reading in the course of learning to read: Zoccolotti et al., 2005.

2. Longitudinal brain imaging of the acquisition of reading: Dehaene-Lambertz et al., 2018.

3. Contribution of parietal cortex to expert reading, only for degraded words: Cohen, Dehaene, Vinckier, Jobert, and Montavont, 2008; Vinckier et al., 2006.

4. Visual recognition of frequent combinations of letters: Binder, Medler, Westbury, Liebenthal, and Buchanan, 2006; Dehaene, Cohen, Sigman, and Vinckier, 2005; Grainger and Whitney, 2004; Vinckier et al., 2007.

5. Tuning of early visual cortex to letter perception: Chang et al., 2015; Dehaene et al., 2010; Sigman et al., 2005; Szwed et al., 2011, 2014.

6. Unconscious reading: Dehaene et al., 2001, 2004.

7. Automatization of arithmetic: Ansari and Dhital, 2006; Rivera, Reiss, Eckert, and Menon, 2005. The hippocampus also seems to strongly contribute to the memory for arithmetic facts: Qin et al., 2014.

8. Sleep interrupts the forgetting curve: Jenkins and Dallenbach, 1924.

9. REM sleep improves learning: Karni, Tanne, Rubenstein, Askenasy, and Sagi, 1994.

10. Sleep and the consolidation of recent learning: Huber, Ghilardi, Massimini, and Tononi, 2004; Stickgold, 2005; Walker, Brakefield, Hobson, and Stickgold, 2003; Walker and Stickgold, 2004.

11. Overexpression of the zif-268 gene during sleep: Ribeiro, Goyal, Mello, and Pavlides, 1999.

12. Neuronal replay during the night: Ji and Wilson, 2007; Louie and Wilson, 2001; Skaggs and McNaughton, 1996; Wilson and McNaughton, 1994.

13. Decoding brain activity during sleep: Chen and Wilson, 2017; Horikawa, Tamaki, Miyawaki, and Kamitani, 2013.

14. Theories of the memory function of sleep: Diekelmann and Born, 2010.

15. Replay during sleep facilitates memory consolidation: Ramanathan, Gulati, and Ganguly, 2015; see also Norimoto et al., 2018, for the direct effect of sleep on synaptic plasticity.

16. Cortical and hippocampal reactivation during sleep in humans: Horikawa et al., 2013; Jiang et al., 2017; Peigneux et al., 2004.

17. Increased slow wave sleep and post-sleep performance improvement: Huber et al., 2004.

18. Brain imaging of the effects of sleep on motor learning: Walker, Stickgold, Alsop, Gaab, and Schlaug, 2005.

19. Boosting slow oscillations during sleep improves memory: Marshall, Helgadóttir, Mölle, and Born, 2006; Ngo, Martinetz, Born, and Mölle, 2013.

20. Odors can bias memory consolidation during sleep: Rasch, Büchel, Gais, and Born, 2007.

21. Sounds can bias replay during sleep and improve subsequent memory: Antony, Gobel, O'Hare, Reber, and Paller, 2012; Bendor and Wilson, 2012; Rudoy, Voss, Westerberg, and Paller, 2009.

22. No learning of novel facts during sleep: Bruce et al., 1970; Emmons and Simon, 1956. Nevertheless, a very recent study suggests that during sleep, we may be able to learn the association between a tone and a smell (Arzi et al., 2012).

23. Gazsi, M. (2018, June 8). Philippe Starck: "I couldn't care less about my life." *The Guardian*, theguardian.com.

24. Mathematical insight during sleep: Wagner, Gais, Haider, Verleger, and Born, 2004.

25. Sleep-wake learning algorithms: Hinton, Dayan, Frey, and Neal, 1995; Hinton, Osindero, and Teh, 2006.

26. Hypothesis that the memory function of sleep may be more efficient in humans: Samson and Nunn, 2015.

27. Greater efficiency of sleep in children than in adults: Wilhelm et al., 2013.

28. Babies generalize word meanings after sleeping: Friedrich, Wilhelm, Born, and Friederici, 2015; Seehagen, Konrad, Herbert, and Schneider, 2015.

29. Positive effect of naps in preschoolers: Kurdziel, Duclos, and Spencer, 2013.

30. Sleep deficits and attention disorders: Avior et al., 2004; Cortese et al., 2013; Hiscock et al., 2015; Prehn-Kristensen et al., 2014.

31. Beneficial effects of delaying school start times for adolescents: American Academy of Pediatrics, 2014; Dunster et al., 2018.

CONCLUSION: RECONCILING EDUCATION WITH NEUROSCIENCE

1. Artificial intelligence inspired by neuroscience and cognitive science: Hassabis, Kumaran, Summerfield, and Botvinick, 2017; Lake et al., 2017.

2. See PISA (Program for International Student Assessment, oecd.org/pisa-fr), TIMSS (Trends in International Mathematics and Science Study), and PIRLS (Progress in International Reading Literacy Study, timssandpirls.bc.edu).

BIBLIOGRAPHY

Abboud, S., Maidenbaum, S., Dehaene, S., and Amedi, A. (2015). A number-form area in the blind. *Nature Communications, 6,* 6026.

Adibpour, P., Dubois, J., and Dehaene-Lambertz, G. (2018). Right but not left hemispheric discrimination of faces in infancy. *Nature Human Behaviour, 2*(1), 67–79.

Ahissar, M., and Hochstein, S. (1993). Attentional control of early perceptual learning. *Proceedings of the National Academy of Sciences, 90*(12), 5718–5722.

Almas, A. N., Degnan, K. A., Radulescu, A., Nelson, C. A., Zeanah, C. H., and Fox, N. A. (2012). Effects of early intervention and the moderating effects of brain activity on institutionalized children's social skills at age 8. *Proceedings of the National Academy of Sciences, 109 Suppl 2,* 17228–17231.

Amalric, M., and Dehaene, S. (2016). Origins of the brain networks for advanced mathematics in expert mathematicians. *Proceedings of the National Academy of Sciences, 113*(18), 4909–4917.

Amalric, M., and Dehaene, S. (2017). Cortical circuits for mathematical knowledge: Evidence for a major subdivision within the brain's semantic networks. *Philosophical Transactions of the Royal Society B: Biological Sciences, 373*(1740), 20160515.

Amalric, M., Denghien, I., and Dehaene, S. (2017). On the role of visual experience in mathematical development: Evidence from blind mathematicians. *Developmental Cognitive Neuroscience, 30,* 314–323.

Amalric, M., Wang, L., Pica, P., Figueira, S., Sigman, M., and Dehaene, S. (2017). The language of geometry: Fast comprehension of geometrical primitives and rules in human adults and preschoolers. *PLOS Computational Biology, 13*(1), e1005273.

Amedi, A., Raz, N., Pianka, P., Malach, R., and Zohary, E. (2003). Early 'visual' cortex activation correlates with superior verbal memory performance in the blind. *Nature Neuroscience*, *6*(7), 758–766.

American Academy of Pediatrics. (2014). School start times for adolescents. *Pediatrics*, *134*(3), 642–649.

Amunts, K., Lenzen, M., Friederici, A. D., Schleicher, A., Morosan, P., Palomero-Gallagher, N., and Zilles, K. (2010). Broca's region: Novel organizational principles and multiple receptor mapping. *PLOS Biology*, *8*(9), e1000489.

Amunts, K., and Zilles, K. (2015). Architectonic mapping of the human brain beyond Brodmann. *Neuron*, *88*(6), 1086–1107.

Anderson, R. C., Wilson, P. T., and Fielding, L. G. (1988). Growth in reading and how children spend their time outside of school. *Reading Research Quarterly*, *23*(3), 285–303.

Ansari, D., and Dhital, B. (2006). Age-related changes in the activation of the intraparietal sulcus during nonsymbolic magnitude processing: An event-related functional magnetic resonance imaging study. *Journal of Cognitive Neuroscience*, *18*(11), 1820–1828.

Antony, J. W., Gobel, E. W., O'Hare, J. K., Reber, P. J., and Paller, K. A. (2012). Cued memory reactivation during sleep influences skill learning. *Nature Neuroscience*, *15*(8), 1114–1116.

Arnould, L. (1900). *Une âme en prison: Histoire de l'éducation d'une aveugle-sourde-muette de naissance*. Paris: Oudin.

Arzi, A., Shedlesky, L., Ben-Shaul, M., Nasser, K., Oksenberg, A., Hairston, I. S., and Sobel, N. (2012). Humans can learn new information during sleep. *Nature Neuroscience*, *15*(10), 1460–1465.

Ashcraft, M. H. (2002). Math anxiety: Personal, educational, and cognitive consequences. *Current Directions in Psychological Science*, *11*(5), 181–185.

Au, J., Sheehan, E., Tsai, N., Duncan, G. J., Buschkuehl, M., and Jaeggi, S. M. (2015). Improving fluid intelligence with training on working memory: A meta-analysis. *Psychonomic Bulletin and Review*, *22*(2), 366–377.

Auble, P. M., and Franks, J. J. (1978). The effects of effort toward comprehension on recall. *Memory and Cognition*, *6*(1), 20–25.

Auble, P. M., Franks, J. J., and Soraci, S. A. (1979). Effort toward comprehension: Elaboration or "aha"? *Memory and Cognition*, *7*(6), 426–434.

Avior, G., Fishman, G., Leor, A., Sivan, Y., Kaysar, N., and Derowe, A. (2004). The effect of tonsillectomy and adenoidectomy on inattention and impulsivity as measured by the Test of Variables of Attention (TOVA) in children with obstructive sleep apnea syndrome. *Otolaryngology*, *131*(4), 367–371.

Bahdanau, D., Cho, K., and Bengio, Y. (2014). Neural machine translation by jointly learning to align and translate. arxiv.org/abs/1409.0473.

Baillargeon, R., and DeVos, J. (1991). Object permanence in young infants: Further evidence. *Child Development*, *62*(6), 1227–1246.

Baillargeon, R., Needham, A., and DeVos, J. (1992). The development of young infants' intuitions about support. *Early Development and Parenting*, 1(2), 69–78.

Baldwin, D. A., Markman, E. M., Bill, B., Desjardins, R. N., Irwin, J. M., and Tidball, G. (1996). Infants' reliance on a social criterion for establishing word-object relations. *Child Development*, 67(6), 3135–3153.

Balsam, P. D., and Gallistel, C. R. (2009). Temporal maps and informativeness in associative learning. *Trends in Neurosciences*, 32(2), 73–78.

Banino, A., Barry, C., Uria, B., Blundell, C., Lillicrap, T., Mirowski, P., . . . Kumaran, D. (2018). Vector-based navigation using grid-like representations in artificial agents. *Nature*, 557(7705), 429–433.

Bao, S., Chan, V. T., and Merzenich, M. M. (2001). Cortical remodelling induced by activity of ventral tegmental dopamine neurons. *Nature*, 412(6842), 79–83.

Bavelier, D., Green, C. S., Han, D. H., Renshaw, P. F., Merzenich, M. M., and Gentile, D. A. (2011). Brains on video games. *Nature Reviews Neuroscience*, 12(12), 763–768.

Beckers, T., Miller, R. R., De Houwer, J., and Urushihara, K. (2006). Reasoning rats: Forward blocking in Pavlovian animal conditioning is sensitive to constraints of causal inference. *Journal of Experimental Psychology: General*, 135(1), 92–102.

Bedny, M. (2017). Evidence from blindness for a cognitively pluripotent cortex. *Trends in Cognitive Sciences*, 21(9), 637–648.

Bedny, M., Pascual-Leone, A., Dodell-Feder, D., Fedorenko, E., and Saxe, R. (2011). Language processing in the occipital cortex of congenitally blind adults. *Proceedings of the National Academy of Sciences*, 108(11), 4429–4434.

Behne, T., Carpenter, M., Call, J., and Tomasello, M. (2005). Unwilling versus unable: Infants' understanding of intentional action. *Developmental Psychology*, 41(2), 328–337.

Bekinschtein, T. A., Dehaene, S., Rohaut, B., Tadel, F., Cohen, L., and Naccache, L. (2009). Neural signature of the conscious processing of auditory regularities. *Proceedings of the National Academy of Sciences*, 106(5), 1672–1677.

Belle, M., Godefroy, D., Couly, G., Malone, S. A., Collier, F., Giacobini, P., and Chédotal, A. (2017). Tridimensional visualization and analysis of early human development. *Cell*, 169(1), 161–173.

Bendor, D., and Wilson, M. A. (2012). Biasing the content of hippocampal replay during sleep. *Nature Neuroscience*, 15(10), 1439–1444.

Berens, A. E., and Nelson, C. A. (2015). The science of early adversity: Is there a role for large institutions in the care of vulnerable children? *Lancet*, 386(9991), 388–398.

Bergman-Nutley, S., and Klingberg, T. (2014). Effect of working memory training on working memory, arithmetic and following instructions. *Psychological Research*, 78(6), 869–877.

Berkes, P., Orbán, G., Lengyel, M., and Fiser, J. (2011). Spontaneous cortical activity reveals hallmarks of an optimal internal model of the environment. *Science*, 331(6013), 83–87.

Bermudez, P., Lerch, J. P., Evans, A. C., and Zatorre, R. J. (2009). Neuroanatomical correlates of musicianship as revealed by cortical thickness and voxel-based morphometry. *Cerebral Cortex, 19*(7), 1583–1596.

Bernal, S., Dehaene-Lambertz, G., Millotte, S., and Christophe, A. (2010). Two-year-olds
compute syntactic structure on-line. *Developmental Science, 13*(1), 69–76.

Bessa, C., Maciel, P., and Rodrigues, A. J. (2013). Using *C. elegans* to decipher the cellular
and molecular mechanisms underlying neurodevelopmental disorders. *Molecular
Neurobiology, 48*(3), 465–489.

Bevins, R. A. (2001). Novelty seeking and reward: Implications for the study of high-risk
behaviors. *Current Directions in Psychological Science, 10*(6), 189–193.

Bialystok, E., Craik, F. I. M., Green, D. W., and Gollan, T. H. (2009). Bilingual minds.
Psychological Science in the Public Interest, 10(3), 89–129.

Binder, J. R., Medler, D. A., Westbury, C. F., Liebenthal, E., and Buchanan, L. (2006).
Tuning of the human left fusiform gyrus to sublexical orthographic structure. *NeuroImage, 33*(2), 739–748.

Blair, C., and Raver, C. C. (2014). Closing the achievement gap through modification of
neurocognitive and neuroendocrine function: Results from a cluster randomized
controlled trial of an innovative approach to the education of children in kindergarten. *PLOS ONE, 9*(11), e112393.

Blair, K. P., Rosenberg-Lee, M., Tsang, J. M., Schwartz, D. L., and Menon, V. (2012).
Beyond natural numbers: Negative number representation in parietal cortex. *Frontiers in Human Neuroscience, 6,* 7.

Bliss, T. V., and Lømo, T. (1973). Long-lasting potentiation of synaptic transmission in
the dentate area of the anaesthetized rabbit following stimulation of the perforant
path. *Journal of Physiology, 232*(2), 331–356.

Bock, A. S., Binda, P., Benson, N. C., Bridge, H., Watkins, K. E., and Fine, I. (2015).
Resting-state retinotopic organization in the absence of retinal input and visual
experience. *Journal of Neuroscience, 35*(36), 12366–12382.

Bonawitz, E., Shafto, P., Gweon, H., Goodman, N. D., Spelke, E., and Schulz, L. (2011).
The double-edged sword of pedagogy: Instruction limits spontaneous exploration
and discovery. *Cognition, 120*(3), 322–330.

Bond, R., and Smith, P. B. (1996). Culture and conformity: A meta-analysis of studies using Asch's (1952b, 1956) line judgment task. *Psychological Bulletin, 119*(1),
111–137.

Borst, G., Poirel, N., Pineau, A., Cassotti, M., and Houdé, O. (2013). Inhibitory control efficiency in a Piaget-like class-inclusion task in school-age children and
adults: A developmental negative priming study. *Developmental Psychology, 49*(7),
1366–1374.

Bouhali, F., Thiebaut de Schotten, M., Pinel, P., Poupon, C., Mangin, J.-F., Dehaene, S.,
and Cohen, L. (2014). Anatomical connections of the visual word form area. *Journal
of Neuroscience, 34*(46), 15402–15414.

Bradley, M. M., Costa, V. D., Ferrari, V., Codispoti, M., Fitzsimmons, J. R., and Lang, P. J. (2015). Imaging distributed and massed repetitions of natural scenes: Spontaneous retrieval and maintenance. *Human Brain Mapping, 36*(4), 1381–1392.

Braga, L. W., Amemiya, E., Tauil, A., Sugueida, D., Lacerda, C., Klein, E., . . . Dehaene, S. (2017). Tracking adult literacy acquisition with functional MRI: A single-case study. *Mind, Brain, and Education, 11*(3), 121–132.

Brewer, J. B., Zhao, Z., Desmond, J. E., Glover, G. H., and Gabrieli, J. D. (1998). Making memories: Brain activity that predicts how well visual experience will be remembered. *Science, 281*(5380), 1185–1187.

Brodmann, K. (1909). *Vergleichende Lokalisationslehre der Grosshirnrinde [Localisation in the cerebral cortex].* Leipzig: Barth.

Bromberg-Martin, E. S., and Hikosaka, O. (2009). Midbrain dopamine neurons signal preference for advance information about upcoming rewards. *Neuron, 63*(1), 119–126.

Bruce, D. J., Evans, C. R., Fenwick, P. B. C., and Spencer, V. (1970). Effect of Presenting Novel Verbal Material during Slow-wave Sleep. *Nature, 225*(5235), 873.

Brun, V. H., Leutgeb, S., Wu, H.-Q., Schwarcz, R., Witter, M. P., Moser, E. I., and Moser, M.-B. (2008). Impaired spatial representation in CA1 after lesion of direct input from entorhinal cortex. *Neuron, 57*(2), 290–302.

Buon, M., Jacob, P., Margules, S., Brunet, I., Dutat, M., Cabrol, D., and Dupoux, E. (2014). Friend or foe? Early social evaluation of human interactions. *PLOS ONE, 9*(2), e88612.

Butler, A. C., Karpicke, J. D., and Roediger, H. L. (2008). Correcting a metacognitive error: Feedback increases retention of low-confidence correct responses. *Journal of Experimental Psychology: Learning, Memory, and Cognition, 34*(4), 918–928.

Butterworth, B. (2010). Foundational numerical capacities and the origins of dyscalculia. *Trends in Cognitive Sciences, 14*(12), 534–541.

Byers-Heinlein, K., and Werker, J. F. (2009). Monolingual, bilingual, trilingual: Infants' language experience influences the development of a word-learning heuristic. *Developmental Science, 12*(5), 815–823.

Callan, D. E., and Schweighofer, N. (2010). Neural correlates of the spacing effect in explicit verbal semantic encoding support the deficient-processing theory. *Human Brain Mapping, 31*(4), 645–659.

Campbell, F. A., Pungello, E. P., Burchinal, M., Kainz, K., Pan, Y., Wasik, B. H., . . . Ramey, C. T. (2012). Adult outcomes as a function of an early childhood educational program: An Abecedarian Project follow-up. *Developmental Psychology, 48*(4), 1033–1043.

Campbell, F., Conti, G., Heckman, J. J., Moon, S. H., Pinto, R., Pungello, E., and Pan, Y. (2014). Early childhood investments substantially boost adult health. *Science, 343*(6178), 1478–1485.

Cantlon, J. F., Brannon, E. M., Carter, E. J., and Pelphrey, K. A. (2006). Functional imaging of numerical processing in adults and 4-y-old children. *PLOS Biology, 4*(5), e125.

Cantlon, J. F., and Li, R. (2013). Neural activity during natural viewing of *Sesame Street* statistically predicts test scores in early childhood. *PLOS Biology, 11*(1), e1001462.

Cardoso-Leite, P., and Bavelier, D. (2014). Video game play, attention, and learning: How to shape the development of attention and influence learning? *Current Opinion in Neurology, 27*(2), 185–191.

Carey, S., and Bartlett, E. (1978). Acquiring a single new word. *Papers and Reports on Child Language Development, 15*, 17–29.

Caroni, P., Donato, F., and Muller, D. (2012). Structural plasticity upon learning: Regulation and functions. *Nature Reviews Neuroscience, 13*(7), 478–490.

Carreiras, M., Seghier, M. L., Baquero, S., Estevez, A., Lozano, A., Devlin, J. T., and Price, C. J. (2009). An anatomical signature for literacy. *Nature, 461*(7266), 983–986.

Carrier, M., and Pashler, H. (1992). The influence of retrieval on retention. *Memory and Cognition, 20*(6), 633–642.

Castles, A., Rastle, K., and Nation, K. (2018). Ending the reading wars: Reading acquisition from novice to expert. *Psychological Science in the Public Interest, 19*(1), 5–51.

Castro-Caldas, A., Petersson, K. M., Reis, A., Stone-Elander, S., and Ingvar, M. (1998). The illiterate brain: Learning to read and write during childhood influences the functional organization of the adult brain. *Brain, 121*(6), 1053–1063.

Cepeda, N. J., Coburn, N., Rohrer, D., Wixted, J. T., Mozer, M. C., and Pashler, H. (2009). Optimizing distributed practice: Theoretical analysis and practical implications. *Experimental Psychology, 56*(4), 236–246.

Cepeda, N. J., Pashler, H., Vul, E., Wixted, J. T., and Rohrer, D. (2006). Distributed practice in verbal recall tasks: A review and quantitative synthesis. *Psychological Bulletin, 132*(3), 354–380.

Cesana-Arlotti, N., Martín, A., Téglás, E., Vorobyova, L., Cetnarski, R., and Bonatti, L. L. (2018). Precursors of logical reasoning in preverbal human infants. *Science, 359*(6381), 1263–1266.

Chafee, M. V. (2013). A scalar neural code for categories in parietal cortex: Representing cognitive variables as "more" or "less." *Neuron, 77*(1), 7–9.

Chakraborty, M., and Jarvis, E. D. (2015). Brain evolution by brain pathway duplication. *Philosophical Transactions of the Royal Society B: Biological Sciences, 370*(1684), 20150056.

Chang, C. H. C., Pallier, C., Wu, D. H., Nakamura, K., Jobert, A., Kuo, W.-J., and Dehaene, S. (2015). Adaptation of the human visual system to the statistics of letters and line configurations. *NeuroImage, 120*, 428–440.

Chao, Z. C., Takaura, K., Wang, L., Fujii, N., and Dehaene, S. (2018). Large-scale cortical networks for hierarchical prediction and prediction error in the primate brain. *Neuron, 100*(5), 1252–1266.

Chen, Z., and Wilson, M. A. (2017). Deciphering neural codes of memory during sleep. *Trends in Neurosciences, 40*(5), 260–275.

Chiao, J. Y. (2010). Neural basis of social status hierarchy across species. *Current Opinion in Neurobiology, 20*(6), 803–809.

Cho, K., Courville, A., and Bengio, Y. (2015). Describing multimedia content using attention-based encoder-decoder networks. *IEEE Transactions on Multimedia*, *17*(11), 1875–1886.

Chun, M. M., and Marois, R. (2002). The dark side of visual attention. *Current Opinion in Neurobiology*, *12*(2), 184–189.

Clark, E. V. (1988). On the logic of contrast. *Journal of Child Language*, *15*(2), 317–335.

Claro, S., Paunesku, D., and Dweck, C. S. (2016). Growth mindset tempers the effects of poverty on academic achievement. *Proceedings of the National Academy of Sciences*, *113*(31), 8664–8668.

Cohen, L., Dehaene, S., McCormick, S., Durant, S., and Zanker, J. M. (2016). Brain mechanisms of recovery from pure alexia: A single case study with multiple longitudinal scans. *Neuropsychologia*, *91*, 36–49.

Cohen, L., Dehaene, S., Vinckier, F., Jobert, A., and Montavont, A. (2008). Reading normal and degraded words: Contribution of the dorsal and ventral visual pathways. *NeuroImage*, *40*(1), 353–366.

Conel, J. L. (1939–67). *The postnatal development of the human cerebral cortex* (Vols. 1–8). Cambridge, MA: Harvard University Press.

Constantinescu, A. O., O'Reilly, J. X., and Behrens, T. E. J. (2016). Organizing conceptual knowledge in humans with a gridlike code. *Science*, *352*(6292), 1464–1468.

Corallo, G., Sackur, J., Dehaene, S., and Sigman, M. (2008). Limits on introspection: Distorted subjective time during the dual-task bottleneck. *Psychological Science*, *19*(11), 1110–1117.

Cortese, S., Brown, T. E., Corkum, P., Gruber, R., O'Brien, L. M., Stein, M., . . . Owens, J. (2013). Assessment and management of sleep problems in youths with attention-deficit/hyperactivity disorder. *Journal of the American Academy of Child and Adolescent Psychiatry*, *52*(8), 784–796.

Costa, A., and Sebastián-Gallés, N. (2014). How does the bilingual experience sculpt the brain? *Nature Reviews Neuroscience*, *15*(5), 336–345.

Courchesne, E., Pierce, K., Schumann, C. M., Redcay, E., Buckwalter, J. A., Kennedy, D. P., and Morgan, J. (2007). Mapping early brain development in autism. *Neuron*, *56*(2), 399–413.

Courtney, S. M., Ungerleider, L. G., Keil, K., and Haxby, J. V. (1997). Transient and sustained activity in a distributed neural system for human working memory. *Nature*, *386*(6625), 608–611.

Craik, F. I. M., and Tulving, E. (1975). Depth of processing and the retention of words in episodic memory. *Journal of Experimental Psychology: General*, *104*(3), 268–294.

Csibra, G., and Gergely, G. (2009). Natural pedagogy. *Trends in Cognitive Sciences*, *13*(4), 148–153.

Çukur, T., Nishimoto, S., Huth, A. G., and Gallant, J. L. (2013). Attention during natural vision warps semantic representation across the human brain. *Nature Neuroscience*, *16*(6), 763–770.

Curran, T., Tucker, D. M., Kutas, M., and Posner, M. I. (1993). Topography of the N400: Brain electrical activity reflecting semantic expectancy. *Electroencephalography and Clinical Neurophysiology, 88*(3), 188–209.

Cyr, M., and Shi, R. (2013). Development of abstract grammatical categorization in infants. *Child Development, 84*(2), 617–629.

Darki, F., Peyrard-Janvid, M., Matsson, H., Kere, J., and Klingberg, T. (2012). Three dyslexia susceptibility genes, DYX1C1, DCDC2, and KIAA0319, affect temporo-parietal white matter structure. *Biological Psychiatry, 72*(8), 671–676.

Deen, B., Richardson, H., Dilks, D. D., Takahashi, A., Keil, B., Wald, L. L., . . . Saxe, R. (2017). Organization of high-level visual cortex in human infants. *Nature Communications, 8*, 13995.

Dehaene, S. (2003). The neural basis of the Weber-Fechner law: A logarithmic mental number line. *Trends in Cognitive Sciences, 7*(4), 145–147.

Dehaene, S. (2005). Evolution of human cortical circuits for reading and arithmetic: The "neuronal recycling" hypothesis. In S. Dehaene, J.-R. Duhamel, M. D. Hauser, and G. Rizzolatti (Eds.), *From monkey brain to human brain* (pp. 133–157). Cambridge, MA: MIT Press.

Dehaene, S. (2007). Symbols and quantities in parietal cortex: Elements of a mathematical theory of number representation and manipulation. In P. Haggard, Y. Rossetti, and M. Kawato (Eds.), *Attention and performance XXII: Sensorimotor foundations of higher cognition* (pp. 527–574). Cambridge, MA: Harvard University Press.

Dehaene, S. (2009). *Reading in the brain: The new science of how we read.* New York, NY: Penguin Group.

Dehaene, S. (2011). *The number sense: How the mind creates mathematics* (2nd ed.). New York, NY: Oxford University Press.

Dehaene, S. (2014). *Consciousness and the brain.* New York, NY: Penguin Group.

Dehaene, S., and Akhavein, R. (1995). Attention, automaticity, and levels of representation in number processing. *Journal of Experimental Psychology: Learning, Memory, and Cognition, 21*(2), 314–326.

Dehaene, S., Bossini, S., and Giraux, P. (1993). The mental representation of parity and numerical magnitude. *Journal of Experimental Psychology: General, 122*(3), 371–396.

Dehaene, S., and Changeux, J. P. (2011). Experimental and theoretical approaches to conscious processing. *Neuron, 70*(2), 200–227.

Dehaene, S., Changeux, J. P., Naccache, L., Sackur, J., and Sergent, C. (2006). Conscious, preconscious, and subliminal processing: A testable taxonomy. *Trends in Cognitive Sciences, 10*(5), 204–211.

Dehaene, S., and Cohen, L. (2007). Cultural recycling of cortical maps. *Neuron, 56*(2), 384–398.

Dehaene, S., Cohen, L., Morais, J., and Kolinsky, R. (2015). Illiterate to literate: Behavioural and cerebral changes induced by reading acquisition. *Nature Reviews Neuroscience, 16*(4), 234–244.

Dehaene, S., Cohen, L., Sigman, M., and Vinckier, F. (2005). The neural code for written words: A proposal. *Trends in Cognitive Sciences*, *9*(7), 335–341.

Dehaene, S., Dupoux, E., and Mehler, J. (1990). Is numerical comparison digital? Analogical and symbolic effects in two-digit number comparison. *Journal of Experimental Psychology: Human Perception and Performance*, *16*(3), 626–641.

Dehaene, S., Izard, V., Pica, P., and Spelke, E. (2006). Core knowledge of geometry in an Amazonian indigene group. *Science*, *311*(5759), 381–384.

Dehaene, S., Izard, V., Spelke, E., and Pica, P. (2008). Log or linear? Distinct intuitions of the number scale in Western and Amazonian indigene cultures. *Science*, *320*(5880), 1217–1220.

Dehaene, S., Jobert, A., Naccache, L., Ciuciu, P., Poline, J.-B., Le Bihan, D., and Cohen, L. (2004). Letter binding and invariant recognition of masked words: Behavioral and neuroimaging evidence. *Psychological Science*, *15*(5), 307–313.

Dehaene, S., Kerszberg, M., and Changeux, J. P. (1998). A neuronal model of a global workspace in effortful cognitive tasks. *Proceedings of the National Academy of Sciences*, *95*(24), 14529–14534.

Dehaene, S., Lau, H., and Kouider, S. (2017). What is consciousness, and could machines have it? *Science*, *358*(6362), 486–492.

Dehaene, S., and Marques, J. F. (2002). Cognitive euroscience: Scalar variability in price estimation and the cognitive consequences of switching to the euro. *Quarterly Journal of Experimental Psychology*, *55*(3), 705–731.

Dehaene, S., Meyniel, F., Wacongne, C., Wang, L., and Pallier, C. (2015). The neural representation of sequences: From transition probabilities to algebraic patterns and linguistic trees. *Neuron*, *88*(1), 2–19.

Dehaene, S., and Naccache, L. (2001). Towards a cognitive neuroscience of consciousness: Basic evidence and a workspace framework. *Cognition*, *79*(1–2), 1–37.

Dehaene, S., Naccache, L., Cohen, L., Le Bihan, D., Mangin, J.-F., Poline, J.-B., and Rivière, D. (2001). Cerebral mechanisms of word masking and unconscious repetition priming. *Nature Neuroscience*, *4*(7), 752–758.

Dehaene, S., Pegado, F., Braga, L. W., Ventura, P., Nunes Filho, G., Jobert, A., ... Cohen, L. (2010). How learning to read changes the cortical networks for vision and language. *Science*, *330*(6009), 1359–1364.

Dehaene-Lambertz, G., Dehaene, S., and Hertz-Pannier, L. (2002). Functional neuroimaging of speech perception in infants. *Science*, *298*(5600), 2013–2015.

Dehaene-Lambertz, G., Hertz-Pannier, L., Dubois, J., Meriaux, S., Roche, A., Sigman, M., and Dehaene, S. (2006). Functional organization of perisylvian activation during presentation of sentences in preverbal infants. *Proceedings of the National Academy of Sciences*, *103*(38), 14240–14245.

Dehaene-Lambertz, G., Monzalvo, K., and Dehaene, S. (2018). The emergence of the visual word form: Longitudinal evolution of category-specific ventral visual areas during reading acquisition. *PLOS Biology*, *16*(3), e2004103.

Dehaene-Lambertz, G., and Spelke, E. S. (2015). The infancy of the human brain. *Neuron, 88*(1), 93–109.

de Lavilléon, G., Lacroix, M. M., Rondi-Reig, L., and Benchenane, K. (2015). Explicit memory creation during sleep demonstrates a causal role of place cells in navigation. *Natura Neuroscience, 18*(4), 493–495.

Denison, S., and Xu, F. (2010). Integrating physical constraints in statistical inference by 11-month-old infants. *Cognitive Science, 34*(5), 885–908.

Dennett, D. C. (1995). *Darwin's dangerous idea: Evolution and the meanings of life.* New York, NY: Simon and Schuster.

Desimone, R., and Duncan, J. (1995). Neural mechanisms of selective visual attention. *Annual Review of Neuroscience, 18*, 193–222.

D'Esposito, M., and Grossman, M. (1996). The physiological basis of executive function and working memory. *Neuroscientist, 2*(6), 345–352.

Diamond, A., and Doar, B. (1989). The performance of human infants on a measure of frontal cortex function, the delayed response task. *Developmental Psychobiology, 22*(3), 271–294.

Diamond, A., and Goldman-Rakic, P. S. (1989). Comparison of human infants and rhesus monkeys on Piaget's AB task: Evidence for dependence on dorsolateral prefrontal cortex. *Experimental Brain Research, 74*(1), 24–40.

Diamond, A., and Lee, K. (2011). Interventions shown to aid executive function development in children 4 to 12 years old. *Science, 333*(6045), 959–964.

Diekelmann, S., and Born, J. (2010). The memory function of sleep. *Nature Reviews Neuroscience, 11*(2), 114–126.

Diester, I., and Nieder, A. (2007). Semantic associations between signs and numerical categories in the prefrontal cortex. *PLOS Biology, 5*(11), e294.

Diester, I., and Nieder, A. (2010). Numerical values leave a semantic imprint on associated signs in monkeys. *Journal of Cognitive Neuroscience, 22*(1), 174–183.

Ditz, H. M., and Nieder, A. (2015). Neurons selective to the number of visual items in the corvid songbird endbrain. *Proceedings of the National Academy of Sciences, 112*(25), 7827–7832.

Doeller, C. F., Barry, C., and Burgess, N. (2010). Evidence for grid cells in a human memory network. *Nature, 463*(7281), 657–661.

Donato, F., Rompani, S. B., and Caroni, P. (2013). Parvalbumin-expressing basket-cell network plasticity induced by experience regulates adult learning. *Nature, 504*(7479), 272–276.

Draganski, B., Gaser, C., Busch, V., Schuierer, G., Bogdahn, U., and May, A. (2004). Neuroplasticity: Changes in grey matter induced by training. *Nature, 427*(6972), 311–312.

Dubois, J., Dehaene-Lambertz, G., Perrin, M., Mangin, J.-F., Cointepas, Y., Duchesnay, E., ... Hertz-Pannier, L. (2007). Asynchrony of the early maturation of white matter bundles in healthy infants: Quantitative landmarks revealed noninvasively by diffusion tensor imaging. *Human Brain Mapping, 29,* 14–27.

Dubois, J., Hertz-Pannier, L., Cachia, A., Mangin, J.-F., Le Bihan, D., and Dehaene-Lambertz, G. (2009). Structural asymmetries in the infant language and sensori-motor networks. *Cerebral Cortex, 19*(2), 414–423.

Dubois, J., Poupon, C., Thirion, B., Simonnet, H., Kulikova, S., Leroy, F., ... Dehaene-Lambertz, G. (2015). Exploring the early organization and maturation of linguistic pathways in the human infant brain. *Cerebral Cortex, 26*(5), 2283–2298.

Dumontheil, I., and Klingberg, T. (2011). Brain activity during a visuospatial working memory task predicts arithmetical performance 2 years later. *Cerebral Cortex, 22*(5), 1078–1085.

Duncan, J. (2003). Intelligence tests predict brain response to demanding task events. *Nature Neuroscience, 6*(3), 207–208.

Duncan, J. (2010). The multiple-demand (MD) system of the primate brain: Mental programs for intelligent behaviour. *Trends in Cognitive Sciences, 14*(4), 172–179.

Duncan, J. (2013). The structure of cognition: Attentional episodes in mind and brain. *Neuron, 80*(1), 35–50.

Dundas, E. M., Plaut, D. C., and Behrmann, M. (2013). The joint development of hemispheric lateralization for words and faces. *Journal of Experimental Psychology: General, 142*(2), 348–358.

Dunlosky, J., Rawson, K. A., Marsh, E. J., Nathan, M. J., and Willingham, D. T. (2013). Improving students' learning with effective learning techniques: Promising directions from cognitive and educational psychology. *Psychological Science in the Public Interest, 14*(1), 4–58.

Dunster, G. P., Iglesia, L. de la, Ben-Hamo, M., Nave, C., Fleischer, J. G., Panda, S., and Iglesia, H. O. de la. (2018). Sleepmore in Seattle: Later school start times are associated with more sleep and better performance in high school students. *Science Advances, 4*(12), eaau6200.

Duyme, M., Dumaret, A.-C., and Tomkiewicz, S. (1999). How can we boost IQs of "dull children"? A late adoption study. *Proceedings of the National Academy of Sciences, 96*(15), 8790–8794.

Dweck, C. S. (2006). *Mindset: The new psychology of success.* New York, NY: Random House.

Egyed, K., Király, I., and Gergely, G. (2013). Communicating shared knowledge in infancy. *Psychological Science, 24*(7), 1348–1353.

Ehri, L. C., Nunes, S. R., Stahl, S. A., and Willows, D. M. (2001). Systematic phonics instruction helps students learn to read: Evidence from the National Reading Panel's meta-analysis. *Review of Educational Research, 71*(3), 393–447.

Ellis, A. W., and Lambon Ralph, M. A. (2000). Age of acquisition effects in adult lexical processing reflect loss of plasticity in maturing systems: Insights from connectionist networks. *Journal of Experimental Psychology: Learning, Memory, and Cognition, 26*(5), 1103–1123.

Elman, J. L., Bates, E. A., Johnson, M. H., Karmiloff-Smith, A., Parisi, D., and Plunkett, K. (1996). *Rethinking innateness: A connectionist perspective on development.* Cambridge, MA: MIT Press.

Elsayed, G. F., Shankar, S., Cheung, B., Papernot, N., Kurakin, A., Goodfellow, I., and Sohl-Dickstein, J. (2018). Adversarial examples that fool both human and computer vision. https://arxiv.org/abs/1802.08195v1.

Elston, G. N. (2003). Cortex, cognition and the cell: New insights into the pyramidal neuron and prefrontal function. *Cerebral Cortex, 13*(11), 1124–1138.

Emmons, W. H., and Simon, C. W. (1956). The non-recall of material presented during sleep. *The American Journal of Psychology, 69,* 76–81.

Epelbaum, M., Milleret, C., Buisseret, P., and Duffer, J. L. (1993). The sensitive period for strabismic amblyopia in humans. *Ophthalmology, 100*(3), 323–327.

Esseily, R., Rat-Fischer, L., Somogyi, E., O'Regan, K. J., and Fagard, J. (2016). Humour production may enhance observational learning of a new tool-use action in 18-month-old infants. *Cognition and Emotion, 30*(4), 817–825.

Ester, E. F., Sprague, T. C., and Serences, J. T. (2015). Parietal and frontal cortex encode stimulus-specific mnemonic representations during visual working memory. *Neuron, 87*(4), 893–905.

Everaert, M. B. H., Huybregts, M. A. C., Chomsky, N., Berwick, R. C., and Bolhuis, J. J. (2015). Structures, not strings: Linguistics as part of the cognitive sciences. *Trends in Cognitive Sciences, 19*(12), 729–743.

Fanselow, M. S. (1998). Pavlovian conditioning, negative feedback, and blocking: Mechanisms that regulate association formation. *Neuron, 20*(4), 625–627.

Fattal, I., Friedmann, N., and Fattal-Valevski, A. (2011). The crucial role of thiamine in the development of syntax and lexical retrieval: A study of infantile thiamine deficiency. *Brain, 134*(6), 1720–1739.

Fawcett, S. L., Wang, Y.-Z., and Birch, E. E. (2005). The critical period for susceptibility of human stereopsis. *Investigative Ophthalmology and Visual Science, 46*(2), 521–525.

Fischer, M. H. (2003). Cognitive representation of negative numbers. *Psychological Science, 14*(3), 278–282.

Fisher, A. V., Godwin, K. E., and Seltman, H. (2014). Visual environment, attention allocation, and learning in young children when too much of a good thing may be bad. *Psychological Science, 25*(7), 1362–1370.

Fitzgerald, J. K., Freedman, D. J., Fanini, A., Bennur, S., Gold, J. I., and Assad, J. A. (2013). Biased associative representations in parietal cortex. *Neuron, 77*(1), 180–191.

Fitzsimonds, R. M., Song, H.-J., and Poo, M.-M. (1997). Propagation of activity-dependent synaptic depression in simple neural networks. *Nature, 388*(6641), 439–448.

Flechsig, P. (1876). *Die Leitungsbahnen im Gehirn und Rückenmark des Menschen auf Grund Entwickelungsgeschichtlicher Untersuchungen.* Leipzig: Engelmann.

Flege, J. E., Munro, M. J., and MacKay, I. R. (1995). Factors affecting strength of perceived foreign accent in a second language. *Journal of the Acoustical Society of America*, 97(5), 3125–3134.

Fleming, S. M., Weil, R. S., Nagy, Z., Dolan, R. J., and Rees, G. (2010). Relating introspective accuracy to individual differences in brain structure. *Science*, 329(5998), 1541–1543.

Fodor, J. A. (1975). *The language of thought*. New York, NY: Thomas Y. Crowell.

Fodor, J. A., and Pylyshyn, Z. W. (1988). Connectionism and cognitive architecture: A critical analysis. *Cognition*, 28(1–2), 3–71.

Fodor, J., and McLaughlin, B. P. (1990). Connectionism and the problem of systematicity: Why Smolensky's solution doesn't work. *Cognition*, 35(2), 183–204.

Frank, M. C., Everett, D. L., Fedorenko, E., and Gibson, E. (2008). Number as a cognitive technology: Evidence from Pirahã language and cognition. *Cognition*, 108(3), 819–824.

Freeman, S., Eddy, S. L., McDonough, M., Smith, M. K., Okoroafor, N., Jordt, H., and Wenderoth, M. P. (2014). Active learning increases student performance in science, engineering, and mathematics. *Proceedings of the National Academy of Sciences*, 111(23), 8410–8415.

Friederici, A. D. (2002). Towards a neural basis of auditory sentence processing. *Trends in Cognitive Sciences*, 6(2), 78–84.

Friedmann, N., Kerbel, N., and Shvimer, L. (2010). Developmental attentional dyslexia. *Cortex*, 46(10), 1216–1237.

Friedmann, N., and Rusou, D. (2015). Critical period for first language: The crucial role of language input during the first year of life. *Current Opinion in Neurobiology*, 35, 27–34.

Friedrich, M., Wilhelm, I., Born, J., and Friederici, A. D. (2015). Generalization of word meanings during infant sleep. *Nature Communications*, 6, 6004.

Friston, K. (2005). A theory of cortical responses. *Philosophical Transactions of the Royal Society B: Biological Sciences*, 360(1456), 815–836.

Froemke, R. C., Merzenich, M. M., and Schreiner, C. E. (2007). A synaptic memory trace for cortical receptive field plasticity. *Nature*, 450(7168), 425–429.

Fukuchi-Shimogori, T., and Grove, E. A. (2001). Neocortex patterning by the secreted signaling molecule FGF8. *Science*, 294(5544), 1071–1074.

Fyhn, M., Molden, S., Witter, M. P., Moser, E. I., and Moser, M.-B. (2004). Spatial representation in the entorhinal cortex. *Science*, 305(5688), 1258–1264.

Galaburda, A. M., LoTurco, J., Ramus, F., Fitch, R. H., and Rosen, G. D. (2006). From genes to behavior in developmental dyslexia. *Nature Neuroscience*, 9(10), 1213–1217.

Galgali, A. R., and Mante, V. (2018). Set in one's thoughts. *Nature Neuroscience*, 21(4), 459–460.

Gallistel, C. R. (1990). *The organization of learning*. Cambridge, MA: MIT Press.

Gaser, C., and Schlaug, G. (2003). Brain structures differ between musicians and non-musicians. *Journal of Neuroscience*, 23(27), 9240–9245.

Gathercole, S. E., Pickering, S. J., Knight, C., and Stegmann, Z. (2004). Working memory skills and educational attainment: Evidence from national curriculum assessments at 7 and 14 years of age. *Applied Cognitive Psychology, 18*(1), 1–16.

Geary, D. C. (2011). Cognitive predictors of achievement growth in mathematics: A five-year longitudinal study. *Developmental Psychology, 47*(6), 1539–1552.

Genzel, L., Rossato, J. I., Jacobse, J., Grieves, R. M., Spooner, P. A., Battaglia, F. P., . . . Morris, R. G. M. (2017). The yin and yang of memory consolidation: Hippocampal and neocortical. *PLOS Biology, 15*(1), e2000531.

George, D., Lehrach, W., Kansky, K., Lázaro-Gredilla, M., Laan, C., Marthi, B., . . . Phoenix, D. S. (2017). A generative vision model that trains with high data efficiency and breaks text-based CAPTCHAs. *Science, 358*(6368).

Gerber, P., Schlaffke, L., Heba, S., Greenlee, M. W., Schultz, T., and Schmidt-Wilcke, T. (2014). Juggling revisited—a voxel-based morphometry study with expert jugglers. *NeuroImage, 95*, 320–325.

Gergely, G., Bekkering, H., and Király, I. (2002). Rational imitation in preverbal infants. *Nature, 415*(6873), 755.

Gergely, G., and Csibra, G. (2003). Teleological reasoning in infancy: The naïve theory of rational action. *Trends in Cognitive Sciences, 7*(7), 287–292.

Gerhand, S., and Barry, C. (1999). Age of acquisition, word frequency, and the role of phonology in the lexical decision task. *Memory and Cognition, 27*(4), 592–602.

Gertler, P., Heckman, J., Pinto, R., Zanolini, A., Vermeersch, C., Walker, S., . . . Grantham-McGregor, S. (2014). Labor market returns to an early childhood stimulation intervention in Jamaica. *Science, 344*(6187), 998–1001.

Glass, A. L., and Kang, M. (2018). Dividing attention in the classroom reduces exam performance. *Educational Psychology, 39*(3), 395–408.

Goldman-Rakic, P. S. (1995). Cellular basis of working memory. *Neuron, 14*(3), 477–485.

Golestani, N., Molko, N., Dehaene, S., Le Bihan, D., and Pallier, C. (2007). Brain structure predicts the learning of foreign speech sounds. *Cerebral Cortex, 17*(3), 575–582.

Golub, M. D., Sadtler, P. T., Oby, E. R., Quick, K. M., Ryu, S. I., Tyler-Kabara, E. C., . . . Yu, B. M. (2018). Learning by neural reassociation. *Nature Neuroscience, 21*(4), 607–616.

Goodfellow, I. J., Pouget-Abadie, J., Mirza, M., Xu, B., Warde-Farley, D., Ozair, S., . . . Bengio, Y. (2014). Generative adversarial networks. arxiv.org/abs/1406.2661.

Goodman, C. S., and Shatz, C. J. (1993). Developmental mechanisms that generate precise patterns of neuronal connectivity. *Cell, 72 Suppl*, 77–98.

Goodman, N. D., Ullman, T. D., and Tenenbaum, J. B. (2011). Learning a theory of causality. *Psychological Review, 118*(1), 110–119.

Gopnik, A., Glymour, C., Sobel, D. M., Schulz, L. E., Kushnir, T., and Danks, D. (2004). A theory of causal learning in children: Causal maps and Bayes nets. *Psychological Review, 111*(1), 3–32.

Gopnik, A., Meltzoff, A. N., and Kuhl, P. K. (1999). *The scientist in the crib: What early learning tells us about the mind*. New York, NY: William Morrow.

Gottlieb, J., Oudeyer, P.-Y., Lopes, M., and Baranes, A. (2013). Information-seeking, curiosity, and attention: Computational and neural mechanisms. *Trends in Cognitive Sciences, 17*(11), 585–593.

Goupil, L., Romand-Monnier, M., and Kouider, S. (2016). Infants ask for help when they know they don't know. *Proceedings of the National Academy of Sciences, 113*(13), 3492–3496.

Grainger, J., and Whitney, C. (2004). Does the huamn mnid raed wrods as a wlohe? *Trends in Cognitive Sciences, 8*(2), 58–59.

Grantham-McGregor, S. M., Powell, C. A., Walker, S. P., and Himes, J. H. (1991). Nutritional supplementation, psychosocial stimulation, and mental development of stunted children: The Jamaican Study. *Lancet, 338*(8758), 1–5.

Green, C. S., and Bavelier, D. (2003). Action video game modifies visual selective attention. *Nature, 423*(6939), 534–537.

Groen, G. J., and Parkman, J. M. (1972). A chronometric analysis of simple addition. *Psychological Review, 79*(4), 329–343.

Grothendieck, A. (1986). *Récoltes et semailles: Réflexions et témoignage sur un passé de mathématicien.* quarante-deux.org/archives/klein/prefaces/Romans_1965-1969/Recoltes_et _semailles.pdf.

Gruber, M. J., Gelman, B. D., and Ranganath, C. (2014). States of curiosity modulate hippocampus-dependent learning via the dopaminergic circuit. *Neuron, 84*(2), 486–496.

Guerguiev, J., Lillicrap, T. P., and Richards, B. A. (2017). Towards deep learning with segregated dendrites. *ELife, 6,* e22901.

Gullick, M. M., and Wolford, G. (2013). Understanding less than nothing: Children's neural response to negative numbers shifts across age and accuracy. *Frontiers in Psychology, 4,* 584.

Gweon, H., Tenenbaum, J. B., and Schulz, L. E. (2010). Infants consider both the sample and the sampling process in inductive generalization. *Proceedings of the National Academy of Sciences, 107*(20), 9066–9071.

Habibi, A., Damasio, A., Ilari, B., Elliott Sachs, M., and Damasio, H. (2018). Music training and child development: A review of recent findings from a longitudinal study. *Annals of the New York Academy of Sciences.*

Hafting, T., Fyhn, M., Molden, S., Moser, M.-B., and Moser, E. I. (2005). Microstructure of a spatial map in the entorhinal cortex. *Nature, 436*(7052), 801–806.

Hahne, A., and Friederici, A. D. (1999). Electrophysiological evidence for two steps in syntactic analysis: Early automatic and late controlled processes. *Journal of Cognitive Neuroscience, 11*(2), 194–205.

Halberda, J., and Feigenson, L. (2008). Developmental change in the acuity of the "number sense": The approximate number system in 3-, 4-, 5-, and 6-year-olds and adults. *Developmental Psychology, 44*(5), 1457–1465.

Hannagan, T., Amedi, A., Cohen, L., Dehaene-Lambertz, G., and Dehaene, S. (2015). Origins of the specialization for letters and numbers in ventral occipitotemporal cortex. *Trends in Cognitive Sciences, 19*(7), 374–382.

Hannagan, T., Nieder, A., Viswanathan, P., and Dehaene, S. (2017). A random-matrix theory of the number sense. *Philosophical Transactions of the Royal Society B: Biological Sciences, 373*(1740), 20170253.

Hartshorne, J. K., Tenenbaum, J. B., and Pinker, S. (2018). A critical period for second language acquisition: Evidence from ⅔ million English speakers. *Cognition, 177,* 263–277.

Hassabis, D., Kumaran, D., Summerfield, C., and Botvinick, M. (2017). Neuroscience-inspired artificial intelligence. *Neuron, 95*(2), 245–258.

Hattie, J. (2008). *Visible learning.* London and New York, NY: Routledge.

Hattie, J. (2017). *L'apprentissage visible pour les enseignants: Connaître son impact pour maximiser le rendement des élèves.* Québec: Presses de l'Université du Québec.

Hauser, M. D., Chomsky, N., and Fitch, W. T. (2002). The faculty of language: What is it, who has it, and how did it evolve? *Science, 298*(5598), 1569–1579.

Hauser, M. D., and Watumull, J. (2017). The Universal Generative Faculty: The source of our expressive power in language, mathematics, morality, and music. *Journal of Neurolinguistics, 43 Part B,* 78–94.

Hay, J. F., Pelucchi, B., Graf Estes, K., and Saffran, J. R. (2011). Linking sounds to meanings: Infant statistical learning in a natural language. *Cognitive Psychology, 63*(2), 93–106.

Heckman, J. J., Moon, S. H., Pinto, R., Savelyev, P. A., and Yavitz, A. (2010). The rate of return to the HighScope Perry Preschool Program. *Journal of Public Economics, 94*(1), 114–128.

Heilbron, M., and Meyniel, F. (2019). Confidence resets reveal hierarchical adaptive learning in humans. *PLOS Computational Biology, 15*(4), e1006972. doi.org/10.1371/journal.pcbi.1006972.

Held, R., and Hein, A. (1963). Movement-produced stimulation in the development of visually guided behavior. *Journal of Comparative and Physiological Psychology, 56*(5), 872–876.

Hensch, T. K. (2005). Critical period plasticity in local cortical circuits. *Nature Reviews Neuroscience, 6*(11), 877–888.

Hespos, S. J., and Baillargeon, R. (2008). Young infants' actions reveal their developing knowledge of support variables: Converging evidence for violation-of-expectation findings. *Cognition, 107*(1), 304–316.

Hinton, G. E., Dayan, P., Frey, B. J., and Neal, R. M. (1995). The "wake-sleep" algorithm for unsupervised neural networks. *Science, 268*(5214), 1158–1161.

Hinton, G. E., Osindero, S., and Teh, Y.-W. (2006). A fast learning algorithm for deep belief nets. *Neural Computation, 18*(7), 1527–1554.

Hiscock, H., Sciberras, E., Mensah, F., Gerner, B., Efron, D., Khano, S., and Oberklaid, F. (2015). Impact of a behavioural sleep intervention on symptoms and sleep in children with attention deficit hyperactivity disorder, and parental mental health: Randomised controlled trial. *BMJ (Clinical Research Ed.), 350,* h68.

Hoeft, F., McCandliss, B. D., Black, J. M., Gantman, A., Zakerani, N., Hulme, C., . . . Gabrieli, J. D. E. (2011). Neural systems predicting long-term outcome in dyslexia. *Proceedings of the National Academy of Sciences, 108*(1), 361–366.

Holtmaat, A., and Caroni, P. (2016). Functional and structural underpinnings of neuronal assembly formation in learning. *Nature Neuroscience, 19*(12), 1553–1562.

Horikawa, T., Tamaki, M., Miyawaki, Y., and Kamitani, Y. (2013). Neural decoding of visual imagery during sleep. *Science, 340*(6132), 639–642.

Houdé, O., Zago, L., Mellet, E., Moutier, S., Pineau, A., Mazoyer, B., and Tzourio-Mazoyer, N. (2000). Shifting from the perceptual brain to the logical brain: The neural impact of cognitive inhibition training. *Journal of Cognitive Neuroscience, 12*(5), 721–728.

Huber, R., Ghilardi, M. F., Massimini, M., and Tononi, G. (2004). Local sleep and learning. *Nature, 430*(6995), 78–81.

Hurley, M. M., Dennett, D. C., and Adams, R. B. (2011). *Inside jokes: Using humor to reverse-engineer the mind.* Cambridge, MA: MIT Press.

Huttenlocher, P. R., and Dabholkar, A. S. (1997). Regional differences in synaptogenesis in human cerebral cortex. *Journal of Comparative Neurology, 387*(2), 167–178.

Hutton, J. S., Horowitz-Kraus, T., Mendelsohn, A. L., DeWitt, T., Holland, S. K., and C-MIND Authorship Consortium. (2015). Home reading environment and brain activation in preschool children listening to stories. *Pediatrics, 136*(3), 466–478.

Hutton, J. S., Phelan, K., Horowitz-Kraus, T., Dudley, J., Altaye, M., DeWitt, T., and Holland, S. K. (2017). Shared reading quality and brain activation during story listening in preschool-age children. *Journal of Pediatrics, 191*, 204–211.e1.

Iriki, A. (2005). A prototype of *Homo faber*: A silent precursor of human intelligence in the tool-using monkey brain. In S. Dehaene, J.-R. Duhamel, M. D. Hauser, and G. Rizzolatti (Eds.), *From monkey brain to human brain* (pp. 253–271). Cambridge, MA: MIT Press.

Isaacs, E. B., Edmonds, C. J., Lucas, A., and Gadian, D. G. (2001). Calculation difficulties in children of very low birthweight: A neural correlate. *Brain, 124*(9), 1701–1707.

Isingrini, M., Perrotin, A., and Souchay, C. (2008). Aging, metamemory regulation and executive functioning. *Progress in Brain Research, 169*, 377–392.

Iuculano, T. (2016). Neurocognitive accounts of developmental dyscalculia and its remediation. *Progress in Brain Research, 227*, 305–333.

Izard, V., Dehaene-Lambertz, G., and Dehaene, S. (2008). Distinct cerebral pathways for object identity and number in human infants. *PLOS Biology, 6*(2), 275–285.

Izard, V., Sann, C., Spelke, E. S., and Streri, A. (2009). Newborn infants perceive abstract numbers. *Proceedings of the National Academy of Sciences, 106*(25), 10382–10385.

Jacob, S. N., and Nieder, A. (2009). Notation-independent representation of fractions in the human parietal cortex. *Journal of Neuroscience, 29*(14), 4652–4657.

Jacoby, L. L., and Dallas, M. (1981). On the relationship between autobiographical memory and perceptual learning. *Journal of Experimental Psychology: General, 110*(3), 306–340.

Jaeggi, S. M., Buschkuehl, M., Jonides, J., and Shah, P. (2011). Short- and long-term benefits of cognitive training. *Proceedings of the National Academy of Sciences, 108*(25), 10081–10086.

James, C. E., Oechslin, M. S., Van De Ville, D., Hauert, C.-A., Descloux, C., and Lazeyras, F. (2014). Musical training intensity yields opposite effects on grey matter density in cognitive versus sensorimotor networks. *Brain Structure and Function, 219*(1), 353–366.

Jaynes, E. T. (2003). *Probability theory: The logic of science*. Cambridge, MA: Cambridge University Press.

Jenkins, J. G., and Dallenbach, K. M. (1924). Obliviscence during sleep and waking. *American Journal of Psychology, 35*(4), 605–612.

Ji, D., and Wilson, M. A. (2007). Coordinated memory replay in the visual cortex and hippocampus during sleep. *Nature Neuroscience, 10*(1), 100–107.

Jiang, X., Long, T., Cao, W., Li, J., Dehaene, S., and Wang, L. (2018). Production of supra-regular spatial sequences by macaque monkeys. *Current Biology, 28*(12), 1851–1859.

Jiang, X., Shamie, I., Doyle, W. K., Friedman, D., Dugan, P., Devinsky, O., . . . Halgren, E. (2017). Replay of large-scale spatio-temporal patterns from waking during subsequent NREM sleep in human cortex. *Scientific Reports, 7*, 17380.

Jo, J., and Bengio, Y. (2017). Measuring the tendency of CNNs to learn surface statistical regularities. arxiv.org/abs/1711.11561.

Johansson, F., Jirenhed, D.-A., Rasmussen, A., Zucca, R., and Hesslow, G. (2014). Memory trace and timing mechanism localized to cerebellar Purkinje cells. *Proceedings of the National Academy of Sciences, 111*(41), 14930–14934.

Johnson, J. S., and Newport, E. L. (1989). Critical period effects in second language learning: The influence of maturational state on the acquisition of English as a second language. *Cognitive Psychology, 21*(1), 60–99.

Josselyn, S. A., Köhler, S., and Frankland, P. W. (2015). Finding the engram. *Nature Reviews Neuroscience, 16*(9), 521–534.

Kaminski, J., Call, J., and Fischer, J. (2004). Word learning in a domestic dog: Evidence for "fast mapping." *Science, 304*(5677), 1682–1683.

Kang, M. J., Hsu, M., Krajbich, I. M., Loewenstein, G., McClure, S. M., Wang, J. T., and Camerer, C. F. (2009). The wick in the candle of learning: Epistemic curiosity activates reward circuitry and enhances memory. *Psychological Science, 20*(8), 963–973.

Kang, S. H. K., Lindsey, R. V., Mozer, M. C., and Pashler, H. (2014). Retrieval practice over the long term: Should spacing be expanding or equal-interval? *Psychonomic Bulletin and Review, 21*(6), 1544–1550.

Kanjlia, S., Lane, C., Feigenson, L., and Bedny, M. (2016). Absence of visual experience modifies the neural basis of numerical thinking. *Proceedings of the National Academy of Sciences, 113*(40), 11172–11177.

Kano, T., Brockie, P. J., Sassa, T., Fujimoto, H., Kawahara, Y., Iino, Y., ... Maricq, A. V. (2008). Memory in *Caenorhabditis elegans* is mediated by NMDA-type ionotropic glutamate receptors. *Current Biology, 18*(13), 1010–1015.

Kaplan, F., and Oudeyer, P.-Y. (2007). In search of the neural circuits of intrinsic motivation. *Frontiers in Neuroscience, 1*(1), 225–236.

Kapur, S., Craik, F. I., Tulving, E., Wilson, A. A., Houle, S., and Brown, G. M. (1994). Neuroanatomical correlates of encoding in episodic memory: Levels of processing effect. *Proceedings of the National Academy of Sciences, 91*(6), 2008–2011.

Karni, A., Tanne, D., Rubenstein, B. S., Askenasy, J., and Sagi, D. (1994). Dependence on REM sleep of overnight improvement of a perceptual skill. *Science, 265*(5172), 679–682.

Karpicke, J. D., and Roediger, H. L. (2008). The critical importance of retrieval for learning. *Science, 319*(5865), 966–968.

Kastner, S., and Ungerleider, L. G. (2000). Mechanisms of visual attention in the human cortex. *Annual Review of Neuroscience, 23*, 315–341.

Keller, H. (1903). *The story of my life.* New York, NY: Doubleday, Page and Co.

Kellman, P. J., and Spelke, E. S. (1983). Perception of partly occluded objects in infancy. *Cognitive Psychology, 15*(4), 483–524.

Kemp, C., and Tenenbaum, J. B. (2008). The discovery of structural form. *Proceedings of the National Academy of Sciences, 105*(31), 10687–10692.

Kidd, C., Piantadosi, S. T., and Aslin, R. N. (2012). The Goldilocks effect: Human infants allocate attention to visual sequences that are neither too simple nor too complex. *PLOS ONE, 7*(5), e36399.

Kidd, C., Piantadosi, S. T., and Aslin, R. N. (2014). The Goldilocks effect in infant auditory attention. *Child Development, 85*(5), 1795–1804.

Kilgard, M. P., and Merzenich, M. M. (1998). Cortical map reorganization enabled by nucleus basalis activity. *Science, 279*(5357), 1714–1718.

Kim, J. J., and Diamond, D. M. (2002). The stressed hippocampus, synaptic plasticity and lost memories. *Nature Reviews Neuroscience, 3*(6), 453–462.

Kim, W. B., and Cho, J.-H. (2017). Encoding of discriminative fear memory by input-specific LTP in the amygdala. *Neuron, 95*(5), 1129–1146.

Kirschner, P. A., Sweller, J., and Clark, R. E. (2006). Why minimal guidance during instruction does not work: An analysis of the failure of constructivist, discovery, problem-based, experiential, and inquiry-based teaching. *Educational Psychologist, 41*(2), 75–86.

Kirschner, P. A., and van Merriënboer, J. J. G. (2013). Do learners really know best? Urban legends in education. *Educational Psychologist, 48*(3), 169–183.

Kitamura, T., Ogawa, S. K., Roy, D. S., Okuyama, T., Morrissey, M. D., Smith, L. M., ... Tonegawa, S. (2017). Engrams and circuits crucial for systems consolidation of a memory. *Science, 356*(6333), 73–78.

Klingberg, T. (2010). Training and plasticity of working memory. *Trends in Cognitive Sciences, 14*(7), 317–324.

Knops, A., Thirion, B., Hubbard, E. M., Michel, V., and Dehaene, S. (2009). Recruitment of an area involved in eye movements during mental arithmetic. *Science, 324*(5934), 1583–1585.

Knops, A., Viarouge, A., and Dehaene, S. (2009). Dynamic representations underlying symbolic and nonsymbolic calculation: Evidence from the operational momentum effect. *Attention, Perception, and Psychophysics, 71*(4), 803–821.

Knudsen, E. I., and Knudsen, P. F. (1990). Sensitive and critical periods for visual calibration of sound localization by barn owls. *Journal of Neuroscience, 10*(1), 222–232.

Knudsen, E. I., Zheng, W., and DeBello, W. M. (2000). Traces of learning in the auditory localization pathway. *Proceedings of the National Academy of Sciences, 97*(22), 11815–11820.

Koechlin, E., Dehaene, S., and Mehler, J. (1997). Numerical transformations in five-month-old human infants. *Mathematical Cognition, 3*(2), 89–104.

Koechlin, E., Ody, C., and Kouneiher, F. (2003). The architecture of cognitive control in the human prefrontal cortex. *Science, 302*(5648), 1181–1185.

Koepp, M. J., Gunn, R. N., Lawrence, A. D., Cunningham, V. J., Dagher, A., Jones, T., . . . Grasby, P. M. (1998). Evidence for striatal dopamine release during a video game. *Nature, 393*(6682), 266–268.

Kolinsky, R., Morais, J., Content, A., and Cary, L. (1987). Finding parts within figures: A developmental study. *Perception, 16*(3), 399–407.

Kolinsky, R., Verhaeghe, A., Fernandes, T., Mengarda, E. J., Grimm-Cabral, L., and Morais, J. (2011). Enantiomorphy through the looking glass: Literacy effects on mirror-image discrimination. *Journal of Experimental Psychology: General, 140*(2), 210–238.

Kontra, C., Goldin-Meadow, S., and Beilock, S. L. (2012). Embodied learning across the life span. *Topics in Cognitive Science, 4*(4), 731–739.

Kontra, C., Lyons, D. J., Fischer, S. M., and Beilock, S. L. (2015). Physical experience enhances science learning. *Psychological Science, 26*(6), 737–749.

Kouider, S., Stahlhut, C., Gelskov, S. V., Barbosa, L. S., Dutat, M., de Gardelle, V., . . . Dehaene-Lambertz, G. (2013). A neural marker of perceptual consciousness in infants. *Science, 340*(6130), 376–380.

Krause, M. R., Zanos, T. P., Csorba, B. A., Pilly, P. K., Choe, J., Phillips, M. E., . . . Pack, C. C. (2017). Transcranial direct current stimulation facilitates associative learning and alters functional connectivity in the primate brain. *Current Biology, 27*(20), 3086–3096.

Kropff, E., and Treves, A. (2008). The emergence of grid cells: Intelligent design or just adaptation? *Hippocampus, 18*(12), 1256–1269.

Krubitzer, L. (2007). The magnificent compromise: Cortical field evolution in mammals. *Neuron, 56*(2), 201–208.

Kuhl, P. K., Tsao, F. M., and Liu, H. M. (2003). Foreign-language experience in infancy: Effects of short-term exposure and social interaction on phonetic learning. *Proceedings of the National Academy of Sciences, 100*(15), 9096–9101.

Kurdziel, L., Duclos, K., and Spencer, R. M. C. (2013). Sleep spindles in midday naps enhance learning in preschool children. *Proceedings of the National Academy of Sciences, 110*(43), 17267–17272.

Kushnir, T., Xu, F., and Wellman, H. M. (2010). Young children use statistical sampling to infer the preferences of other people. *Psychological Science, 21*(8), 1134–1140.

Kutas, M., and Federmeier, K. D. (2011). Thirty years and counting: Finding meaning in the N400 component of the event-related brain potential (ERP). *Annual Review of Psychology, 62,* 621–647.

Kutas, M., and Hillyard, S. A. (1980). Reading senseless sentences: Brain potentials reflect semantic incongruity. *Science, 207*(4427), 203–205.

Kutter, E. F., Bostroem, J., Elger, C. E., Mormann, F., and Nieder, A. (2018). Single neurons in the human brain encode numbers. *Neuron, 100*(3), 753–761.

Kwan, K. Y., Lam, M. M. S., Johnson, M. B., Dube, U., Shim, S., Rašin, M.-R., ... Šestan, N. (2012). Species-dependent posttranscriptional regulation of NOS1 by FMRP in the developing cerebral cortex. *Cell, 149*(4), 899–911.

Lake, B. M., Salakhutdinov, R., and Tenenbaum, J. B. (2015). Human-level concept learning through probabilistic program induction. *Science, 350*(6266), 1332–1338.

Lake, B. M., Ullman, T. D., Tenenbaum, J. B., and Gershman, S. J. (2017). Building machines that learn and think like people. *Behavioral and Brain Sciences, 40,* e253.

Landau, B., Gleitman, H., and Spelke, E. (1981). Spatial knowledge and geometric representation in a child blind from birth. *Science, 213*(4513), 1275–1278.

Lane, C., Kanjlia, S., Omaki, A., and Bedny, M. (2015). "Visual" cortex of congenitally blind adults responds to syntactic movement. *Journal of Neuroscience, 35*(37), 12859–12868.

Langston, R. F., Ainge, J. A., Couey, J. J., Canto, C. B., Bjerknes, T. L., Witter, M. P., ... Moser, M.-B. (2010). Development of the spatial representation system in the rat. *Science, 328*(5985), 1576–1580.

LeCun, Y., Bengio, Y., and Hinton, G. (2015). Deep learning. *Nature, 521*(7553), 436–444.

LeCun, Y., Bottou, L., Bengio, Y., and Haffner, P. (1998). Gradient-based learning applied to document recognition. *Proceedings of the IEEE, 86*(11), 2278–2324.

Lefevre, J., and Mangin, J.-F. (2010). A reaction-diffusion model of human brain development. *PLOS Computational Biology, 6*(4), e1000749.

Leong, Y. C., Radulescu, A., Daniel, R., DeWoskin, V., and Niv, Y. (2017). Dynamic interaction between reinforcement learning and attention in multidimensional environments. *Neuron, 93*(2), 451–463.

Leppanen, P. H., Richardson, U., Pihko, E., Eklund, K. M., Guttorm, T. K., Aro, M., and Lyytinen, H. (2002). Brain responses to changes in speech sound durations differ between infants with and without familial risk for dyslexia. *Developmental Neuropsychology, 22*(1), 407–422.

Lerner, Y., Honey, C. J., Silbert, L. J., and Hasson, U. (2011). Topographic mapping of a hierarchy of temporal receptive windows using a narrated story. *Journal of Neuroscience, 31*(8), 2906–2915.

Leroy, F., Cai, Q., Bogart, S. L., Dubois, J., Coulon, O., Monzalvo, K., . . . Dehaene-Lambertz, G. (2015). New human-specific brain landmark: The depth asymmetry of superior temporal sulcus. *Proceedings of the National Academy of Sciences, 112*(4), 1208–1213.

Li, P., Legault, J., and Litcofsky, K. A. (2014). Neuroplasticity as a function of second language learning: Anatomical changes in the human brain. *Cortex, 58*, 301–324.

Li, S., Lee, K., Zhao, J., Yang, Z., He, S., and Weng, X. (2013). Neural competition as a developmental process: Early hemispheric specialization for word processing delays specialization for face processing. *Neuropsychologia, 51*(5), 950–959.

Lillard, A., and Else-Quest, N. (2006). Evaluating Montessori education. *Science, 313*(5795), 1893–1894.

Lindsey, R. V., Shroyer, J. D., Pashler, H., and Mozer, M. C. (2014). Improving students' long-term knowledge retention through personalized review. *Psychological Science, 25*(3), 639–647.

Lisman, J., Buzsáki, G., Eichenbaum, H., Nadel, L., Ranganath, C., and Redish, A. D. (2017). Viewpoints: How the hippocampus contributes to memory, navigation and cognition. *Nature Neuroscience, 20*(11), 1434–1447.

Liu, S., Ullman, T. D., Tenenbaum, J. B., and Spelke, E. S. (2017). Ten-month-old infants infer the value of goals from the costs of actions. *Science, 358*(6366), 1038–1041.

Livingstone, M. S., Vincent, J. L., Arcaro, M. J., Srihasam, K., Schade, P. F., and Savage, T. (2017). Development of the macaque face-patch system. *Nature Communications, 8*, 14897.

Loewenstein, G. (1994). The psychology of curiosity: A review and reinterpretation. *Psychological Bulletin, 116*(1), 75–98.

Lømo, T. (2018). Discovering long-term potentiation (LTP)—recollections and reflections on what came after. *Acta Physiologica, 222*(2), e12921.

Louie, K., and Wilson, M. A. (2001). Temporally structured replay of awake hippocampal ensemble activity during rapid eye movement sleep. *Neuron, 29*(1), 145–156.

Lyons, I. M., and Beilock, S. L. (2012). When math hurts: Math anxiety predicts pain network activation in anticipation of doing math. *PLOS ONE, 7*(10), e48076.

Lyons, K. E., and Ghetti, S. (2011). The development of uncertainty monitoring in early childhood. *Child Development, 82*(6), 1778–1787.

Lyytinen, H., Ahonen, T., Eklund, K., Guttorm, T., Kulju, P., Laakso, M. L., . . . Viholainen, H. (2004). Early development of children at familial risk for dyslexia—follow-up from birth to school age. *Dyslexia, 10*(3), 146–178.

Ma, L., and Xu, F. (2013). Preverbal infants infer intentional agents from the perception of regularity. *Developmental Psychology, 49*(7), 1330–1337.

Mack, A., and Rock, I. (1998). *Inattentional blindness.* Cambridge, MA: MIT Press.

Maguire, E. A., Gadian, D. G., Johnsrude, I. S., Good, C. D., Ashburner, J., Frackowiak, R. S., and Frith, C. D. (2000). Navigation-related structural change in the hippocampi of taxi drivers. *Proceedings of the National Academy of Sciences, 97*(8), 4398–4403.

Maguire, E. A., Spiers, H. J., Good, C. D., Hartley, T., Frackowiak, R. S., and Burgess, N. (2003). Navigation expertise and the human hippocampus: A structural brain imaging analysis. *Hippocampus, 13*(2), 250–259.

Mahmoudzadeh, M., Dehaene-Lambertz, G., Fournier, M., Kongolo, G., Goudjil, S., Dubois, J., ... Wallois, F. (2013). Syllabic discrimination in premature human infants prior to complete formation of cortical layers. *Proceedings of the National Academy of Sciences, 110*(12), 4846–4851.

Mahon, B. Z., Anzellotti, S., Schwarzbach, J., Zampini, M., and Caramazza, A. (2009). Category-specific organization in the human brain does not require visual experience. *Neuron, 63*(3), 397–405.

Maloney, E. A., and Beilock, S. L. (2012). Math anxiety: Who has it, why it develops, and how to guard against it. *Trends in Cognitive Sciences, 16*(8), 404–406.

Markman, E. M., and Wachtel, G. F. (1988). Children's use of mutual exclusivity to constrain the meanings of words. *Cognitive Psychology, 20*(2), 121–157.

Markman, E. M., Wasow, J. L., and Hansen, M. B. (2003). Use of the mutual exclusivity assumption by young word learners. *Cognitive Psychology, 47*(3), 241–275.

Marois, R., and Ivanoff, J. (2005). Capacity limits of information processing in the brain. *Trends in Cognitive Sciences, 9*(6), 296–305.

Marques, J. F., and Dehaene, S. (2004). Developing intuition for prices in euros: Rescaling or relearning prices? *Journal of Experimental Psychology: Applied, 10*(3), 148–155.

Marshall, C. (2017). Montessori education: A review of the evidence base. *npj Science of Learning, 2*(1), 11.

Marshall, L., Helgadóttir, H., Mölle, M., and Born, J. (2006). Boosting slow oscillations during sleep potentiates memory. *Nature, 444*(7119), 610–613.

Marti, S., King, J.-R., and Dehaene, S. (2015). Time-resolved decoding of two processing chains during dual-task interference. *Neuron, 88*(6), 1297–1307.

Marti, S., Sigman, M., and Dehaene, S. (2012). A shared cortical bottleneck underlying attentional blink and psychological refractory period. *NeuroImage, 59*(3), 2883–2898.

Martin, S. L., Ramey, C. T., and Ramey, S. (1990). The prevention of intellectual impairment in children of impoverished families: Findings of a randomized trial of educational day care. *American Journal of Public Health, 80*(7), 844–847.

Maye, J., Werker, J. F., and Gerken, L. (2002). Infant sensitivity to distributional information can affect phonetic discrimination. *Cognition, 82*(3), B101–B111.

Mayer, R. E. (2004). Should there be a three-strikes rule against pure discovery learning? The case for guided methods of instruction. *American Psychologist, 59*(1), 14–19.

McCandliss, B. D., Fiez, J. A., Protopapas, A., Conway, M., and McClelland, J. L. (2002). Success and failure in teaching the [r]-[l] contrast to Japanese adults: Tests of a Hebbian model of plasticity and stabilization in spoken language perception. *Cognitive, Affective, and Behavioral Neuroscience, 2*(2), 89–108.

McCloskey, M., and Rapp, B. (2000). A visually based developmental reading deficit. *Journal of Memory and Language, 43*(2), 157–181.

McCrink, K., and Wynn, K. (2004). Large-number addition and subtraction by 9-month-old infants. *Psychological Science, 15*(11), 776–781.

Mehler, J., Jusczyk, P., Lambertz, G., Halsted, N., Bertoncini, J., and Amiel-Tison, C. (1988). A precursor of language acquisition in young infants. *Cognition, 29*(2), 143–178.

Meyer, T., and Olson, C. R. (2011). Statistical learning of visual transitions in monkey inferotemporal cortex. *Proceedings of the National Academy of Sciences, 108*(48), 19401–19406.

Meyniel, F., and Dehaene, S. (2017). Brain networks for confidence weighting and hierarchical inference during probabilistic learning. *Proceedings of the National Academy of Sciences, 114*(19), E3859–E3868.

Millum, J., and Emanuel, E. J. (2007). The ethics of international research with abandoned children. *Science, 318*(5858), 1874–1875.

Mnih, V., Kavukcuoglu, K., Silver, D., Rusu, A. A., Veness, J., Bellemare, M. G., . . . Hassabis, D. (2015). Human-level control through deep reinforcement learning. *Nature, 518*(7540), 529–533.

Mongelli, V., Dehaene, S., Vinckier, F., Peretz, I., Bartolomeo, P., and Cohen, L. (2017). Music and words in the visual cortex: The impact of musical expertise. *Cortex, 86*, 260–274.

Mongillo, G., Barak, O., and Tsodyks, M. (2008). Synaptic theory of working memory. *Science, 319*(5869), 1543–1546.

Monzalvo, K., Fluss, J., Billard, C., Dehaene, S., and Dehaene-Lambertz, G. (2012). Cortical networks for vision and language in dyslexic and normal children of variable socio-economic status. *NeuroImage, 61*(1), 258–274.

Morais, J. (2017). Literacy and democracy. *Language, Cognition and Neuroscience, 33*(3), 351–372.

Morais, J., Bertelson, P., Cary, L., and Alegria, J. (1986). Literacy training and speech segmentation. *Cognition, 24*(1–2), 45–64.

Morais, J., and Kolinsky, R. (2005). Literacy and cognitive change. In M. J. Snowling and C. Hulme (Eds.), *The science of reading: A handbook* (pp. 188–203). Oxford: Blackwell.

Moreno, S., Bialystok, E., Barac, R., Schellenberg, E. G., Cepeda, N. J., and Chau, T. (2011). Short-term music training enhances verbal intelligence and executive function. *Psychological Science, 22*(11), 1425–1433.

Morrison, C. M., and Ellis, A. W. (1995). Roles of word frequency and age of acquisition in word naming and lexical decision. *Journal of Experimental Psychology: Learning, Memory, and Cognition, 21*(1), 116–133.

Morton, J., and Johnson, M. H. (1991). CONSPEC and CONLERN: A two-process theory of infant face recognition. *Psychological Review, 98*(2), 164–181.

Moyer, R. S., and Landauer, T. K. (1967). Time required for judgements of numerical inequality. *Nature, 215*(5109), 1519–1520.

Muckli, L., Naumer, M. J., and Singer, W. (2009). Bilateral visual field maps in a patient with only one hemisphere. *Proceedings of the National Academy of Sciences, 106*(31), 13034–13039.

Musso, M., Moro, A., Glauche, V., Rijntjes, M., Reichenbach, J., Buchel, C., and Weiller, C. (2003). Broca's area and the language instinct. *Nature Neuroscience, 6*(7), 774–781.

Naatanen, R., Paavilainen, P., Rinne, T., and Alho, K. (2007). The mismatch negativity (MMN) in basic research of central auditory processing: A review. *Clinical Neurophysiology, 118*(12), 2544–2590.

Nabokov, V. (1962). *Pale fire*. New York, NY: Putnam.

National Institute of Child Health and Human Development. (2000). Report of the National Reading Panel: Teaching children to read: An evidence-based assessment of the scientific research literature on reading and its implications for reading instruction (NIH publication no. 00-4769). Washington, DC: US Government Printing Office.

Nau, M., Navarro Schröder, T., Bellmund, J. L. S., and Doeller, C. F. (2018). Hexadirectional coding of visual space in human entorhinal cortex. *Nature Neuroscience, 21*(2), 188–190.

Nelson, C. A., Zeanah, C. H., Fox, N. A., Marshall, P. J., Smyke, A. T., and Guthrie, D. (2007). Cognitive recovery in socially deprived young children: The Bucharest Early Intervention Project. *Science, 318*(5858), 1937–1940.

Nelson, M. J., El Karoui, I., Giber, K., Yang, X., Cohen, L., Koopman, H., . . . Dehaene, S. (2017). Neurophysiological dynamics of phrase-structure building during sentence processing. *Proceedings of the National Academy of Sciences, 114*(18), E3669–E3678.

Nemmi, F., Helander, E., Helenius, O., Almeida, R., Hassler, M., Räsänen, P., and Klingberg, T. (2016). Behavior and neuroimaging at baseline predict individual response to combined mathematical and working memory training in children. *Developmental Cognitive Neuroscience, 20*, 43–51.

Ngo, H.-V. V., Martinetz, T., Born, J., and Mölle, M. (2013). Auditory closed-loop stimulation of the sleep slow oscillation enhances memory. *Neuron, 78*(3), 545–553.

Nieder, A., and Dehaene, S. (2009). Representation of number in the brain. *Annual Review of Neuroscience, 32*, 185–208.

Niogi, S. N., and McCandliss, B. D. (2006). Left lateralized white matter microstructure accounts for individual differences in reading ability and disability. *Neuropsychologia, 44*(11), 2178–2188.

Noble, K. G., Norman, M. F., and Farah, M. J. (2005). Neurocognitive correlates of socioeconomic status in kindergarten children. *Developmental Science*, *8*(1), 74–87.

Norimoto, H., Makino, K., Gao, M., Shikano, Y., Okamoto, K., Ishikawa, T., . . . Ikegaya, Y. (2018). Hippocampal ripples down-regulate synapses. *Science*, *359*(6383), 1524–1527.

Obayashi, S., Suhara, T., Kawabe, K., Okauchi, T., Maeda, J., Akine, Y., . . . Iriki, A. (2001). Functional brain mapping of monkey tool use. *NeuroImage*, *14*(4), 853–861.

Oechslin, M. S., Gschwind, M., and James, C. E. (2018). Tracking training-related plasticity by combining fMRI and DTI: The right hemisphere ventral stream mediates musical syntax processing. *Cerebral Cortex*, *28*(4), 1209–1218.

Olah, C., Mordvintsev, A., and Schubert, L. (2017). Feature visualization. *Distill*. doi .org/10.23915/distill.00007.

Olesen, P. J., Westerberg, H., and Klingberg, T. (2004). Increased prefrontal and parietal activity after training of working memory. *Nature Neuroscience*, *7*(1), 75–79.

Orbán, G., Berkes, P., Fiser, J., and Lengyel, M. (2016). Neural variability and sampling-based probabilistic representations in the visual cortex. *Neuron*, *92*(2), 530–543.

Paller, K. A., McCarthy, G., and Wood, C. C. (1988). ERPs predictive of subsequent recall and recognition performance. *Biological Psychology*, *26*(1–3), 269–276.

Pallier, C., Dehaene, S., Poline, J.-B., Le Bihan, D., Argenti, A. M., Dupoux, E., and Mehler, J. (2003). Brain imaging of language plasticity in adopted adults: Can a second language replace the first? *Cerebral Cortex*, *13*(2), 155–161.

Pallier, C., Devauchelle, A. D., and Dehaene, S. (2011). Cortical representation of the constituent structure of sentences. *Proceedings of the National Academy of Sciences*, *108*(6), 2522–2527.

Palminteri, S., Kilford, E. J., Coricelli, G., and Blakemore, S.-J. (2016). The computational development of reinforcement learning during adolescence. *PLOS Computational Biology*, *12*(6), e1004953.

Pashler, H., McDaniel, M., Rohrer, D., and Bjork, R. (2008). Learning styles: Concepts and evidence. *Psychological Science in the Public Interest*, *9*(3), 105–119.

Pegado, F., Comerlato, E., Ventura, F., Jobert, A., Nakamura, K., Buiatti, M., . . . Dehaene, S. (2014). Timing the impact of literacy on visual processing. *Proceedings of the National Academy of Sciences*, *111*(49), E5233–E5242.

Pegado, F., Nakamura, K., Braga, L. W., Ventura, P., Nunes Filho, G., Pallier, C., . . . Dehaene, S. (2014). Literacy breaks mirror invariance for visual stimuli: A behavioral study with adult illiterates. *Journal of Experimental Psychology: General*, *143*(2), 887–894.

Peigneux, P., Laureys, S., Fuchs, S., Collette, F., Perrin, F., Reggers, J., . . . Maquet, P. (2004). Are spatial memories strengthened in the human hippocampus during slow wave sleep? *Neuron*, *44*(3), 535–545.

Pena, M., Werker, J. F., and Dehaene-Lambertz, G. (2012). Earlier speech exposure does not accelerate speech acquisition. *Journal of Neuroscience*, *32*(33), 11159–11163.

Penn, D. C., Holyoak, K. J., and Povinelli, D. J. (2008). Darwin's mistake: Explaining the discontinuity between human and nonhuman minds. *Behavioral and Brain Sciences*, *31*(2), 109–130; discussion 130–178.

Pessiglione, M., Seymour, B., Flandin, G., Dolan, R. J., and Frith, C. D. (2006). Dopamine-dependent prediction errors underpin reward-seeking behaviour in humans. *Nature*, *442*(7106), 1042–1045.

Piantadosi, S. T., Jara-Ettinger, J., and Gibson, E. (2014). Children's learning of number words in an indigenous farming-foraging group. *Developmental Science*, *17*(4), 553–563.

Piantadosi, S. T., Tenenbaum, J. B., and Goodman, N. D. (2012). Bootstrapping in a language of thought: A formal model of numerical concept learning. *Cognition*, *123*(2), 199–217.

Piantadosi, S. T., Tenenbaum, J. B., and Goodman, N. D. (2016). The logical primitives of thought: Empirical foundations for compositional cognitive models. *Psychological Review*, *123*(4), 392–424.

Piazza, M., De Feo, V., Panzeri, S., and Dehaene, S. (2018). Learning to focus on number. *Cognition*, *181*, 35–45.

Piazza, M., Facoetti, A., Trussardi, A. N., Berteletti, I., Conte, S., Lucangeli, D., . . . Zorzi, M. (2010). Developmental trajectory of number acuity reveals a severe impairment in developmental dyscalculia. *Cognition*, *116*(1), 33–41.

Piazza, M., Izard, V., Pinel, P., Le Bihan, D., and Dehaene, S. (2004). Tuning curves for approximate numerosity in the human intraparietal sulcus. *Neuron*, *44*(3), 547–555.

Piazza, M., Pica, P., Izard, V., Spelke, E. S., and Dehaene, S. (2013). Education enhances the acuity of the nonverbal approximate number system. *Psychological Science*, *24*(6), 1037–1043.

Pica, P., Lemer, C., Izard, V., and Dehaene, S. (2004). Exact and approximate arithmetic in an Amazonian indigene group. *Science*, *306*(5695), 499–503.

Pierce, L. J., Klein, D., Chen, J.-K., Delcenserie, A., and Genesee, F. (2014). Mapping the unconscious maintenance of a lost first language. *Proceedings of the National Academy of Sciences*, *111*(48), 17314–17319.

Pinheiro-Chagas, P., Dotan, D., Piazza, M., and Dehaene, S. (2017). Finger tracking reveals the covert stages of mental arithmetic. *Open Mind*, *1*(1), 30–41.

Pittenger, C., and Kandel, E. R. (2003). In search of general mechanisms for long-lasting plasticity: Aplysia and the hippocampus. *Philosophical Transactions of the Royal Society B: Biological Sciences*, *358*(1432), 757–763.

Poirel, N., Borst, G., Simon, G., Rossi, S., Cassotti, M., Pineau, A., and Houdé, O. (2012). Number conservation is related to children's prefrontal inhibitory control: An fMRI study of a Piagetian task. *PLOS ONE*, *7*(7), e40802.

Poo, M.-M., Pignatelli, M., Ryan, T. J., Tonegawa, S., Bonhoeffer, T., Martin, K. C., . . . Stevens, C. (2016). What is memory? The present state of the engram. *BMC Biology*, *14*, 40.

Posner, M. I. (1994). Attention: The mechanisms of consciousness. *Proceedings of the National Academy of Sciences, 91*(16), 7398–7403.

Posner, M. I., and Rothbart, M. K. (1998). Attention, self-regulation and consciousness. *Philosophical Transactions of the Royal Society B: Biological Sciences, 353*(1377), 1915–1927.

Prado, E. L., and Dewey, K. G. (2014). Nutrition and brain development in early life. *Nutrition Reviews, 72*(4), 267–284.

Prehn-Kristensen, A., Munz, M., Göder, R., Wilhelm, I., Korr, K., Vahl, W., ... Baving, L. (2014). Transcranial oscillatory direct current stimulation during sleep improves declarative memory consolidation in children with attention-deficit/hyperactivity disorder to a level comparable to healthy controls. *Brain Stimulation, 7*(6), 793–799.

Qin, S., Cho, S., Chen, T., Rosenberg-Lee, M., Geary, D. C., and Menon, V. (2014). Hippocampal-neocortical functional reorganization underlies children's cognitive development. *Nature Neuroscience, 17*(9), 1263–1269.

Quartz, S. R., and Sejnowski, T. J. (1997). The neural basis of cognitive development: A constructivist manifesto. *Behavioral and Brain Sciences, 20*(4), 537–556; discussion 556–596.

Rakic, P., Bourgeois, J. P., Eckenhoff, M. F., Zecevic, N., and Goldman-Rakic, P. S. (1986). Concurrent overproduction of synapses in diverse regions of the primate cerebral cortex. *Science, 232*(4747), 232–235.

Ramanathan, D. S., Gulati, T., and Ganguly, K. (2015). Sleep-dependent reactivation of ensembles in motor cortex promotes skill consolidation. *PLOS Biology, 13*(9), e1002263.

Ramirez, S., Liu, X., Lin, P.-A., Suh, J., Pignatelli, M., Redondo, R. L., ... Tonegawa, S. (2013). Creating a false memory in the hippocampus. *Science, 341*(6144), 387–391.

Ramirez, S., Liu, X., MacDonald, C. J., Moffa, A., Zhou, J., Redondo, R. L., and Tonegawa, S. (2015). Activating positive memory engrams suppresses depression-like behaviour. *Nature, 522*(7556), 335–339.

Rankin, C. H. (2004). Invertebrate learning: What can't a worm learn? *Current Biology, 14*(15), R617–R618.

Rasch, B., Büchel, C., Gais, S., and Born, J. (2007). Odor cues during slow-wave sleep prompt declarative memory consolidation. *Science, 315*(5817), 1426–1429.

Rasmussen, A., Jirenhed, D. A., and Hesslow, G. (2008). Simple and complex spike firing patterns in Purkinje cells during classical conditioning. *Cerebellum, 7*(4), 563–566.

Rattan, A., Savani, K., Chugh, D., and Dweck, C. S. (2015). Leveraging mindsets to promote academic achievement: Policy recommendations. *Perspectives on Psychological Science, 10*(6), 721–726.

Reich, L., Szwed, M., Cohen, L., and Amedi, A. (2011). A ventral visual stream reading center independent of visual experience. *Current Biology, 21*(5), 363–368.

Reid, V. M., Dunn, K., Young, R. J., Amu, J., Donovan, T., and Reissland, N. (2017). The human fetus preferentially engages with face-like visual stimuli. *Current Biology, 27*(12), 1825–1828.

Rescorla, R. A., and Wagner, A. R. (1972). A theory of Pavlovian conditioning: Variations in the effectiveness of reinforcement and nonreinforcement. In A. H. Black and W.

F. Prokasy (Eds.), *Classical conditioning II: Current research and theory* (pp. 64–99). New York, NY: Appleton-Century-Crofts.

Ribeiro, S., Goyal, V., Mello, C. V., and Pavlides, C. (1999). Brain gene expression during REM sleep depends on prior waking experience. *Learning and Memory, 6*(5), 500–508.

Ritchie, S. J., and Tucker-Drob, E. M. (2018). How much does education improve intelligence? A meta-analysis. *Psychological Science, 29*(8), 1358–1369.

Rivera, S. M., Reiss, A. L., Eckert, M. A., and Menon, V. (2005). Developmental changes in mental arithmetic: Evidence for increased functional specialization in the left inferior parietal cortex. *Cerebral Cortex, 15*(11), 1779–1790.

Robey, A. M., Dougherty, M. R., and Buttaccio, D. R. (2017). Making retrospective confidence judgments improves learners' ability to decide what *not* to study. *Psychological Science, 28*(11), 1683–1693.

Roediger, H. L., and Karpicke, J. D. (2006). Test-enhanced learning: Taking memory tests improves long-term retention. *Psychological Science, 17*(3), 249–255.

Rohrer, D., and Taylor, K. (2006). The effects of overlearning and distributed practise on the retention of mathematics knowledge. *Applied Cognitive Psychology, 20*(9), 1209–1224.

Rohrer, D., and Taylor, K. (2007). The shuffling of mathematics problems improves learning. *Instructional Science, 35*(6), 481–498.

Romeo, R. R., Leonard, J. A., Robinson, S. T., West, M. R., Mackey, A. P., Rowe, M. L., and Gabrieli, J. D. E. (2018). Beyond the 30-million-word gap: Children's conversational exposure is associated with language-related brain function. *Psychological Science, 29*(5), 700–710.

Rouault, M., and Koechlin, E. (2018). Prefrontal function and cognitive control: From action to language. *Current Opinion in Behavioral Sciences, 21*, 106–111.

Rudoy, J. D., Voss, J. L., Westerberg, C. E., and Paller, K. A. (2009). Strengthening individual memories by reactivating them during sleep. *Science, 326*(5956), 1079.

Rueckl, J. G., Paz-Alonso, P. M., Molfese, P. J., Kuo, W.-J., Bick, A., Frost, S. J., . . . Frost, R. (2015). Universal brain signature of proficient reading: Evidence from four contrasting languages. *Proceedings of the National Academy of Sciences, 112*(50), 15510–15515.

Rueda, M. R., Rothbart, M. K., McCandliss, B. D., Saccomanno, L., and Posner, M. I. (2005). Training, maturation, and genetic influences on the development of executive attention. *Proceedings of the National Academy of Sciences, 102*(41), 14931–14936.

Rugani, R., Fontanari, L., Simoni, E., Regolin, L., and Vallortigara, G. (2009). Arithmetic in newborn chicks. *Proceedings of the Royal Society B: Biological Sciences, 276*(1666), 2451–2460.

Rugani, R., Vallortigara, G., Priftis, K., and Regolin, L. (2015). Number-space mapping in the newborn chick resembles humans' mental number line. *Science, 347*(6221), 534–536.

Sabbah, N., Authié, C. N., Sanda, N., Mohand-Saïd, S., Sahel, J.-A., Safran, A. B., . . . Amedi, A. (2016). Increased functional connectivity between language and visually deprived areas in late and partial blindness. *NeuroImage, 136*, 162–173.

Sackur, J., and Dehaene, S. (2009). The cognitive architecture for chaining of two mental operations. *Cognition, 111*(2), 187–211.

Sadtler, P. T., Quick, K. M., Golub, M. D., Chase, S. M., Ryu, S. I., Tyler-Kabara, E. C., . . . Batista, A. P. (2014). Neural constraints on learning. *Nature, 512*(7515), 423–426.

Saffran, J. R., Aslin, R. N., and Newport, E. L. (1996). Statistical learning by 8-month-old infants. *Science, 274*(5294), 1926–1928.

Sakai, T., Mikami, A., Tomonaga, M., Matsui, M., Suzuki, J., Hamada, Y., . . . Matsuzawa, T. (2011). Differential prefrontal white matter development in chimpanzees and humans. *Current Biology, 21*(16), 1397–1402.

Salimpoor, V. N., van den Bosch, I., Kovacevic, N., McIntosh, A. R., Dagher, A., and Zatorre, R. J. (2013). Interactions between the nucleus accumbens and auditory cortices predict music reward value. *Science, 340*(6129), 216–219.

Samson, D. R., and Nunn, C. L. (2015). Sleep intensity and the evolution of human cognition. *Evolutionary Anthropology, 24*(6), 225–237.

Sangrigoli, S., Pallier, C., Argenti, A.-M., Ventureyra, V. A. G., and de Schonen, S. (2005). Reversibility of the other-race effect in face recognition during childhood. *Psychological Science, 16*(6), 440–444.

Saygin, Z. M., Norton, E. S., Osher, D. E., Beach, S. D., Cyr, A. B., Ozernov-Palchik, O., . . . Gabrieli, J. D. E. (2013). Tracking the roots of reading ability: White matter volume and integrity correlate with phonological awareness in prereading and early-reading kindergarten children. *Journal of Neuroscience, 33*(33), 13251–13258.

Saygin, Z. M., Osher, D. E., Koldewyn, K., Reynolds, G., Gabrieli, J. D., and Saxe, R. R. (2012). Anatomical connectivity patterns predict face selectivity in the fusiform gyrus. *Nature Neuroscience, 15*(2), 321–327.

Saygin, Z. M., Osher, D. E., Norton, E. S., Youssoufian, D. A., Beach, S. D., Feather, J., . . . Kanwisher, N. (2016). Connectivity precedes function in the development of the visual word form area. *Nature Neuroscience, 19*(9), 1250–1255.

Schapiro, A. C., Turk-Browne, N. B., Norman, K. A., and Botvinick, M. M. (2016). Statistical learning of temporal community structure in the hippocampus. *Hippocampus, 26*(1), 3–8.

Schlaug, G., Jancke, L., Huang, Y., Staiger, J. F., and Steinmetz, H. (1995). Increased corpus callosum size in musicians. *Neuropsychologia, 33*(8), 1047–1055.

Schmidt, R. A., and Bjork, R. A. (1992). New conceptualizations of practice: Common principles in three paradigms suggest new concepts for training. *Psychological Science, 3*(4), 207–217.

Schoenemann, P. T., Sheehan, M. J., and Glotzer, L. D. (2005). Prefrontal white matter volume is disproportionately larger in humans than in other primates. *Nature Neuroscience, 8*(2), 242–252.

Schultz, W., Dayan, P., and Montague, P. R. (1997). A neural substrate of prediction and reward. *Science, 275*(5306), 1593–1599.

Schweinhart, L. J. (1993). Significant benefits: The High/Scope Perry Preschool study through age 27. Monographs of the High/Scope Educational Research Foundation, no. ten. Education Resources Information Center.

Sederberg, P. B., Kahana, M. J., Howard, M. W., Donner, E. J., and Madsen, J. R. (2003). Theta and gamma oscillations during encoding predict subsequent recall. *Journal of Neuroscience, 23*(34), 10809–10814.

Sederberg, P. B., Schulze-Bonhage, A., Madsen, J. R., Bromfield, E. B., McCarthy, D. C., Brandt, A., . . . Kahana, M. J. (2006). Hippocampal and neocortical gamma oscillations predict memory formation in humans. *Cerebral Cortex, 17*(5), 1190–1196.

Seehagen, S., Konrad, C., Herbert, J. S., and Schneider, S. (2015). Timely sleep facilitates declarative memory consolidation in infants. *Proceedings of the National Academy of Sciences, 112*(5), 1625–1629.

Seitz, A., Lefebvre, C., Watanabe, T., and Jolicoeur, P. (2005). Requirement for high-level processing in subliminal learning. *Current Biology, 15*(18), R753–R755.

Senghas, A., Kita, S., and Özyürek, A. (2004). Children creating core properties of language: Evidence from an emerging sign language in Nicaragua. *Science, 305*(5691), 1779–1782.

Sergent, C., Baillet, S., and Dehaene, S. (2005). Timing of the brain events underlying access to consciousness during the attentional blink. *Nature Neuroscience, 8*(10), 1391–1400.

Shah, P. E., Weeks, H. M., Richards, B., and Kaciroti, N. (2018). Early childhood curiosity and kindergarten reading and math academic achievement. *Pediatric Research, 84*(3), 380–386.

Shatz, C. J. (1996). Emergence of order in visual system development. *Proceedings of the National Academy of Sciences, 93*(2), 602–608.

Shaywitz, S. E., Escobar, M. D., Shaywitz, B. A., Fletcher, J. M., and Makuch, R. (1992). Evidence that dyslexia may represent the lower tail of a normal distribution of reading ability. *New England Journal of Medicine, 326*(3), 145–150.

Sheese, B. E., Rothbart, M. K., Posner, M. I., White, L. K., and Fraundorf, S. H. (2008). Executive attention and self-regulation in infancy. *Infant Behavior and Development, 31*(3), 501–510.

Sheridan, M. A., Fox, N. A., Zeanah, C. H., McLaughlin, K. A., and Nelson, C. A. (2012). Variation in neural development as a result of exposure to institutionalization early in childhood. *Proceedings of the National Academy of Sciences, 109*(32), 12927–12932.

Shi, R., and Lepage, M. (2008). The effect of functional morphemes on word segmentation in preverbal infants. *Developmental Science, 11*(3), 407–413.

Shipston-Sharman, O., Solanka, L., and Nolan, M. F. (2016). Continuous attractor network models of grid cell firing based on excitatory–inhibitory interactions. *Journal of Physiology, 594*(22), 6547–6557.

Shneidman, L. A., Arroyo, M. E., Levine, S. C., and Goldin-Meadow, S. (2013). What counts as effective input for word learning? *Journal of Child Language, 40*(3), 672–686.

Shneidman, L. A., and Goldin-Meadow, S. (2012). Language input and acquisition in a Mayan village: How important is directed speech? *Developmental Science, 15*(5), 659–673.

Shohamy, D., and Turk-Browne, N. B. (2013). Mechanisms for widespread hippocampal involvement in cognition. *Journal of Experimental Psychology: General, 142*(4), 1159–1170.

Siegler, R. S. (1989). Mechanisms of cognitive development. *Annual Review of Psychology, 40*, 353–379.

Siegler, R. S., and Opfer, J. E. (2003). The development of numerical estimation: Evidence for multiple representations of numerical quantity. *Psychological Science, 14*(3), 237–243.

Siegler, R. S., Thompson, C. A., and Schneider, M. (2011). An integrated theory of whole number and fractions development. *Cognitive Psychology, 62*(4), 273–296.

Sigman, M., and Dehaene, S. (2008). Brain mechanisms of serial and parallel processing during dual-task performance. *Journal of Neuroscience, 28*(30), 7585–7598.

Sigman, M., Pan, H., Yang, Y., Stern, E., Silbersweig, D., and Gilbert, C. D. (2005). Top-down reorganization of activity in the visual pathway after learning a shape identification task. *Neuron, 46*(5), 823–835.

Silver, D., Huang, A., Maddison, C. J., Guez, A., Sifre, L., van den Driessche, G., . . . Hassabis, D. (2016). Mastering the game of Go with deep neural networks and tree search. *Nature, 529*(7587), 484–489.

Simons, D. J., and Chabris, C. F. (1999). Gorillas in our midst: Sustained inattentional blindness for dynamic events. *Perception, 28*(9), 1059–1074.

Sisk, V. F., Burgoyne, A. P., Sun, J., Butler, J. L., and Macnamara, B. N. (2018). To what extent and under which circumstances are growth mind-sets important to academic achievement? Two meta-analyses. *Psychological Science, 29*(4), 549–571.

Skaggs, W. E., and McNaughton, B. L. (1996). Replay of neuronal firing sequences in rat hippocampus during sleep following spatial experience. *Science, 271*(5257), 1870–1873.

Smaers, J. B., Gómez-Robles, A., Parks, A. N., and Sherwood, C. C. (2017). Exceptional evolutionary expansion of prefrontal cortex in great apes and humans. *Current Biology, 27*(5), 714–720.

Spelke, E. S. (2003). What makes us smart? Core knowledge and natural language. In D. Gentner and S. Goldin-Meadow (Eds.), *Language in mind: Advances in the study of language and thought* (pp. 277–311). Cambridge, MA: MIT Press.

Spencer, S. J., Steele, C. M., and Quinn, D. M. (1999). Stereotype threat and women's math performance. *Journal of Experimental Social Psychology, 35*(1), 4–28.

Spencer-Smith, M., and Klingberg, T. (2015). Benefits of a working memory training program for inattention in daily life: A systematic review and meta-analysis. *PLOS ONE, 10*(3), e0119522.

Srihasam, K., Mandeville, J. B., Morocz, I. A., Sullivan, K. J., and Livingstone, M. S. (2012). Behavioral and anatomical consequences of early versus late symbol training in macaques. *Neuron, 73*(3), 608–619.

Stahl, A. E., and Feigenson, L. (2015). Observing the unexpected enhances infants' learning and exploration. *Science, 348*(6230), 91–94.

Starkey, P., and Cooper, R. G. (1980). Perception of numbers by human infants. *Science, 210*(4473), 1033–1035.

Starkey, P., Spelke, E. S., and Gelman, R. (1990). Numerical abstraction by human infants. *Cognition, 36*(2), 97–127.

Steele, C. M., and Aronson, J. (1995). Stereotype threat and the intellectual test performance of African Americans. *Journal of Personality and Social Psychology, 69*(5), 797–811.

Steinhauer, K., and Drury, J. E. (2012). On the early left-anterior negativity (ELAN) in syntax studies. *Brain and Language, 120*(2), 135–162.

Stickgold, R. (2005). Sleep-dependent memory consolidation. *Nature, 437*(7063), 1272–1278.

Strauss, M., Sitt, J. D., King, J.-R., Elbaz, M., Azizi, L., Buiatti, M., ... Dehaene, S. (2015). Disruption of hierarchical predictive coding during sleep. *Proceedings of the National Academy of Sciences, 112*(11), E1353–E1362.

Striem-Amit, E., and Amedi, A. (2014). Visual cortex extrastriate body-selective area activation in congenitally blind people "seeing" by using sounds. *Current Biology, 24*(6), 687–692.

Strnad, L., Peelen, M. V., Bedny, M., and Caramazza, A. (2013). Multivoxel pattern analysis reveals auditory motion information in MT+ of both congenitally blind and sighted individuals. *PLOS ONE, 8*(4), e63198.

Sun, T., Patoine, C., Abu-Khalil, A., Visvader, J., Sum, E., Cherry, T. J., ... Walsh, C. A. (2005). Early asymmetry of gene transcription in embryonic human left and right cerebral cortex. *Science, 308*(5729), 1794–1798.

Sun, Z. Y., Klöppel, S., Rivière, D., Perrot, M., Frackowiak, R., Siebner, H., and Mangin, J.-F. (2012). The effect of handedness on the shape of the central sulcus. *NeuroImage, 60*(1), 332–339.

Sur, M., Garraghty, P. E., and Roe, A. W. (1988). Experimentally induced visual projections into auditory thalamus and cortex. *Science, 242*(4884), 1437–1441.

Sur, M., and Rubenstein, J. L. R. (2005). Patterning and plasticity of the cerebral cortex. *Science, 310*(5749), 805–810.

Sutton, R. S., and Barto, A. G. (1998). *Reinforcement learning: An introduction.* Cambridge, MA: MIT Press.

Szpunar, K. K., Khan, N. Y., and Schacter, D. L. (2013). Interpolated memory tests reduce mind wandering and improve learning of online lectures. *Proceedings of the National Academy of Sciences, 110*(16), 6313–6317.

Szwed, M., Dehaene, S., Kleinschmidt, A., Eger, E., Valabregue, R., Amadon, A., and Cohen, L. (2011). Specialization for written words over objects in the visual cortex. *NeuroImage, 56*(1), 330–344.

Szwed, M., Qiao, E., Jobert, A., Dehaene, S., and Cohen, L. (2014). Effects of literacy in early visual and occipitotemporal areas of Chinese and French readers. *Journal of Cognitive Neuroscience, 26*(3), 459–475.

Szwed, M., Ventura, P., Querido, L., Cohen, L., and Dehaene, S. (2012). Reading acquisition enhances an early visual process of contour integration. *Developmental Science*,
   15(1), 139–149.

Takeuchi, T., Duszkiewicz, A. J., and Morris, R. G. M. (2014). The synaptic plasticity
   and memory hypothesis: Encoding, storage and persistence. *Philosophical Transactions
   of the Royal Society B: Biological Sciences*, *369*(1633), 20130288.

Tenenbaum, J. B., Kemp, C., Griffiths, T. L., and Goodman, N. D. (2011). How to grow
   a mind: Statistics, structure, and abstraction. *Science*, *331*(6022), 1279–1285.

Terrace, H. S., Petitto, L. A., Sanders, R. J., and Bever, T. G. (1979). Can an ape create a
   sentence? *Science*, *206*(4421), 891–902.

Thiebaut de Schotten, M., Cohen, L., Amemiya, E., Braga, L. W., and Dehaene, S.
   (2014). Learning to read improves the structure of the arcuate fasciculus. *Cerebral
   Cortex*, *24*(4), 989–995.

Thornton, A., and McAuliffe, K. (2006). Teaching in wild meerkats. *Science*, *313*(5784),
   227–229.

Todorovic, A., and de Lange, F. P. (2012). Repetition suppression and expectation
   suppression are dissociable in time in early auditory evoked fields. *Journal of
   Neuroscience*, *32*(39), 13389–13395.

Tombu, M., and Jolicoeur, P. (2004). Virtually no evidence for virtually perfect time-
   sharing. *Journal of Experimental Psychology: Human Perception and Performance*, *30*(5),
   795–810.

Uhrig, L., Dehaene, S., and Jarraya, B. (2014). A hierarchy of responses to auditory regularities in the macaque brain. *Journal of Neuroscience*, *34*(4), 1127–1132.

van Kerkoerle, T., Self, M. W., and Roelfsema, P. R. (2017). Layer-specificity in the
   effects of attention and working memory on activity in primary visual cortex. *Nature
   Communications*, *8*, 13804.

van Praag, H., Kempermann, G., and Gage, F. H. (2000). Neural consequences of environmental enrichment. *Nature Reviews Neuroscience*, *1*(3), 191–198.

van Vugt, B., Dagnino, B., Vartak, D., Safaai, H., Panzeri, S., Dehaene, S., and Roelfsema, P. R. (2018). The threshold for conscious report: Signal loss and response bias
   in visual and frontal cortex. *Science*, *360*(6388), 537–542.

Ventura, P., Fernandes, T., Cohen, L., Morais, J., Kolinsky, R., and Dehaene, S. (2013).
   Literacy acquisition reduces the influence of automatic holistic processing of faces
   and houses. *Neuroscience Letters*, *554*, 105–109.

Vinckier, F., Dehaene, S., Jobert, A., Dubus, J. P., Sigman, M., and Cohen, L. (2007).
   Hierarchical coding of letter strings in the ventral stream: Dissecting the inner
   organization of the visual word-form system. *Neuron*, *55*(1), 143–156.

Vinckier, F., Naccache, L., Papeix, C., Forget, J., Hahn-Barma, V., Dehaene, S., and Cohen,
   L. (2006). "What" and "where" in word reading: Ventral coding of written words
   revealed by parietal atrophy. *Journal of Cognitive Neuroscience*, *18*(12), 1998–2012.

Viswanathan, P., and Nieder, A. (2013). Neuronal correlates of a visual "sense of number" in primate parietal and prefrontal cortices. *Proceedings of the National Academy of Sciences, 110*(27), 11187–11192.

Viswanathan, P., and Nieder, A. (2015). Differential impact of behavioral relevance on quantity coding in primate frontal and parietal neurons. *Current Biology, 25*(10), 1259–1269.

Vogel, E. K., and Machizawa, M. G. (2004). Neural activity predicts individual differences in visual working memory capacity. *Nature, 428*(6984), 748–751.

Voss, M. W., Vivar, C., Kramer, A. F., and van Praag, H. (2013). Bridging animal and human models of exercise-induced brain plasticity. *Trends in Cognitive Sciences, 17*(10), 525–544.

Wacongne, C., Labyt, E., van Wassenhove, V., Bekinschtein, T., Naccache, L., and Dehaene, S. (2011). Evidence for a hierarchy of predictions and prediction errors in human cortex. *Proceedings of the National Academy of Sciences, 108*(51), 20754–20759.

Waelti, P., Dickinson, A., and Schultz, W. (2001). Dopamine responses comply with basic assumptions of formal learning theory. *Nature, 412*(6842), 43–48.

Wagner, A. D., Schacter, D. L., Rotte, M., Koutstaal, W., Maril, A., Dale, A. M., . . . Buckner, R. L. (1998). Building memories: Remembering and forgetting of verbal experiences as predicted by brain activity. *Science, 281*(5380), 1188–1191.

Wagner, U., Gais, S., Haider, H., Verleger, R., and Born, J. (2004). Sleep inspires insight. *Nature, 427*(6972), 352–355.

Walker, M. P., Brakefield, T., Hobson, J. A., and Stickgold, R. (2003). Dissociable stages of human memory consolidation and reconsolidation. *Nature, 425*(6958), 616–620.

Walker, M. P., and Stickgold, R. (2004). Sleep-dependent learning and memory consolidation. *Neuron, 44*(1), 121–133.

Walker, M. P., Stickgold, R., Alsop, D., Gaab, N., and Schlaug, G. (2005). Sleep-dependent motor memory plasticity in the human brain. *Neuroscience, 133*(4), 911–917.

Walker, S. P., Chang, S. M., Powell, C. A., and Grantham-McGregor, S. M. (2005). Effects of early childhood psychosocial stimulation and nutritional supplementation on cognition and education in growth-stunted Jamaican children: Prospective cohort study. *Lancet, 366*(9499), 1804–1807.

Wang, L., and Krauzlis, R. J. (2018). Visual selective attention in mice. *Current Biology, 28*(5), 676–685.

Wang, L., Uhrig, L., Jarraya, B., and Dehaene, S. (2015). Representation of numerical and sequential patterns in macaque and human brains. *Current Biology, 25*(15), 1966–1974.

Warneken, F., and Tomasello, M. (2006). Altruistic helping in human infants and young chimpanzees. *Science, 311*(5765), 1301–1303.

Watanabe, T., Nanez, J. E., and Sasaki, Y. (2001). Perceptual learning without perception. *Nature, 413*(6858), 844–848.

Weber-Fox, C. M., and Neville, H. J. (1996). Maturational constraints on functional specializations for language processing: ERP and behavioral evidence in bilingual speakers. *Journal of Cognitive Neuroscience, 8*(3), 231–256.

Werker, J. F., and Hensch, T. K. (2014). Critical periods in speech perception: New directions. *Annual Review of Psychology, 66*, 173–196.

Werker, J. F., and Tees, R. C. (1984). Cross-language speech perception: Evidence for perceptual reorganization during the first year of life. *Infant Behavior and Development, 7*(1), 49–63.

Whitlock, J. R., Heynen, A. J., Shuler, M. G., and Bear, M. F. (2006). Learning induces long-term potentiation in the hippocampus. *Science, 313*(5790), 1093–1097.

Widloski, J., and Fiete, I. R. (2014). A model of grid cell development through spatial exploration and spike time-dependent plasticity. *Neuron, 83*(2), 481–495.

Wilhelm, I., Rose, M., Imhof, K. I., Rasch, B., Büchel, C., and Born, J. (2013). The sleeping child outplays the adult's capacity to convert implicit into explicit knowledge. *Nature Neuroscience, 16*(4), 391–393.

Wills, T. J., Cacucci, F., Burgess, N., and O'Keefe, J. (2010). Development of the hippocampal cognitive map in preweanling rats. *Science, 328*(5985), 1573–1576.

Wilson, M. A., and McNaughton, B. L. (1994). Reactivation of hippocampal ensemble memories during sleep. *Science, 265*(5172), 676–679.

Windsor, J., Moraru, A., Nelson, C. A., Fox, N. A., and Zeanah, C. H. (2013). Effect of foster care on language learning at eight years: Findings from the Bucharest Early Intervention Project. *Journal of Child Language, 40*(3), 605–627.

Wynn, K. (1992). Addition and subtraction by human infants. *Nature, 358*, 749–750.

Xu, F., and Garcia, V. (2008). Intuitive statistics by 8-month-old infants. *Proceedings of the National Academy of Sciences, 105*(13), 5012–5015.

Xu, F., and Tenenbaum, J. B. (2007). Word learning as Bayesian inference. *Psychological Review, 114*(2), 245–272.

Xu, K., Ba, J., Kiros, R., Cho, K., Courville, A., Salakhutdinov, R., ... Bengio, Y. (2015). Show, attend and tell: Neural image caption generation with visual attention. arxiv .org/abs/1502.03044.

Yang, C. (2013). Ontogeny and phylogeny of language. *Proceedings of the National Academy of Sciences, 110*(16), 6324–6327.

Yoncheva, Y. N., Blau, V. C., Maurer, U., and McCandliss, B. D. (2010). Attentional focus during learning impacts N170 ERP responses to an artificial script. *Developmental Neuropsychology, 35*(4), 423–445.

Yoon, J. M. D., Johnson, M. H., and Csibra, G. (2008). Communication-induced memory biases in preverbal infants. *Proceedings of the National Academy of Sciences, 105*(36), 13690–13695.

Yoon, K., Buice, M. A., Barry, C., Hayman, R., Burgess, N., and Fiete, I. R. (2013). Specific evidence of low-dimensional continuous attractor dynamics in grid cells. *Nature Neuroscience, 16*(8), 1077–1084.

Young, C. B., Wu, S. S., and Menon, V. (2012). The neurodevelopmental basis of math anxiety. *Psychological Science, 23*(5), 492–501.

Zaromb, F. M., Karpicke, J. D., and Roediger, H. L. (2010). Comprehension as a basis for metacognitive judgments: Effects of effort after meaning on recall and metacognition. *Journal of Experimental Psychology: Learning, Memory, and Cognition, 36*(2), 552–557.

Zaromb, F. M., and Roediger, H. L. I. (2010). The testing effect in free recall is associated with enhanced organizational processes. *Memory and Cognition, 38*(8), 995–1008.

Zhu, X., Wang, F., Hu, H., Sun, X., Kilgard, M. P., Merzenich, M. M., and Zhou, X. (2014). Environmental acoustic enrichment promotes recovery from developmentally degraded auditory cortical processing. *Journal of Neuroscience, 34*(16), 5406–5415.

Zoccolotti, P., De Luca, M., Di Pace, E., Gasperini, F., Judica, A., and Spinelli, D. (2005). Word length effect in early reading and in developmental dyslexia. *Brain and Language, 93*(3), 369–373.

Zylberberg, A., Dehaene, S., Roelfsema, P. R., and Sigman, M. (2011). The human Turing machine: A neural framework for mental programs. *Trends in Cognitive Sciences, 15*(7), 293–300.

INDEX

Page numbers in italics refer to illustrations in the text.

CREDITS

of Sciences of the United States of America, 106(31), 13034–13039, (2009). https://www.pnas.org/content/106/31/13034.

Insert Figure 12: adapted from Amalric, Marie, and Stanislas Dehaene. "Origins of the brain networks for advanced mathematics in expert mathematicians." *Proceedings of the National Academy of Sciences of the United States of America*, vol. 113 (18) 4909-4917, (2016). https://www.pnas.org/content/early/2016/04/06/1603205113 .

Insert Figure 13, bottom: adapted from Amalric, Marie, Isabelle Denghien, and Stanislas Dehaene. "On the role of visual experience in mathematical development: Evidence from blind mathematicians." *Developmental Cognitive Neuroscience*, vol. 30 pg. 314-323 (2018). https://www.sciencedirect.com/science/article/pii/S1878929316302201?via%3Dihub. Licensed under Creative Commons Non-Commercial No-Derivatives 4.0 International License CC BY-NC-ND 4.0. https://creativecommons.org/licenses/by-nc-nd/4.0/.

Insert Figure 14: figure created by the author, from data published in Dehaene, Stanislas, Felipe Pegado, Lucia W. Braga, Paulo Ventura, Gilberto Nunes Filho, Antionette Jobert, Ghislaine Dehaene-Lambertz, Régine Kolinsky, José Morais, and Laurent Cohen. "How Learning to Read Changes the Cortical Networks for Vision and Language." *Science*, vol. 330, issue 6009, pg. 1359–1364, (2010). https://science.sciencemag.org/content/330/6009/1359.

Insert Figure 15, top: courtesy of G. Dehaene-Lambertz.

Insert Figure 15, bottom: figure created by the author from as yet unpublished data and from data published by Monzalvo, Karla, Joel Fluss, Catherine Billard, Stanislas Dehaene, and Ghislaine Dehaene-Lambertz. "Cortical networks for vision and language in dyslexic and normal children of variable socio-economic status." *Neuroimage*, vol. 61(1), pg. 258–274, (2012). https://doi.org/10.1016/j.neuroimage.2012.02.035.

Insert Figure 16, top: adapted from figure by Bruce Blaus, Blausen.com staff. "Medical gallery of Blausen Medical 2014". *WikiJournal of Medicine* 1 (2), (2014). doi:10.15347/wjm/2014.010. Licensed under Creative Commons Attribution-ShareAlike 4.0 International License (CC BY-SA 4.0).

Insert Figure 16, bottom: from Kilgard, Michael P., and Michael M. Merzenich. "Cortical Map Reorganization Enabled by Nucleus Basalis Activity." *Science*, vol. 279, issue 5357, (1998), pg. 1714–8. Reprinted with permission from AAAS.

Insert Figure 17: figure created by the author, from data published in Bekinschtein, Tristan A., Stanislas Dehaene, Benjamin Rohaut, François Tadel, Laurent Cohen, & Lionel Naccache. "Neural signature of the conscious processing of auditory regularities." *Proceedings from the National Academy of Sciences U.S.A.*, vol. 106(5), pg. 1672–1677, (2009). https://doi.org/10.1073/pnas.0809667106; and Strauss, Melanie, Jacobo D. Sitt, Jean-Remi King, Maxime Elbaz, Leila Azizi, Marco Buiatti, Lionel Naccache, Virginia van Wassenhove, and Stanislas Dehaene. "Disruption of hierarchical predictive coding during sleep." *Proceedings of the National Academy of Sciences of the United States of America*, vol. 112(11), E1353-1362, (2015). https://doi.org/10.1073/pnas.1501026112.

Insert Figure 18, left: adapted from Dehaene-Lambertz, Ghislaine, Karla Monzalvo, and Stanislas Dehaene. "The emergence of the visual word form: Longitudinal evolution of category-specific ventral visual areas during reading acquisition." *PLoS Biology*, 16(3), e2004103, (2018). https://journals.plos.org/plosbiology/article?id=10.1371/journal.pbio.2004103. Licensed under Creative Commons Attribution License CC-BY 4.0.

Insert Figure 18, right: redrawn from Zoccolotti, Pierluigi, Maria De Luca, Enrico Di Pace, Filippo Gasperini, Anna Judica, & Donatella Spinelli. "Word length effect in early reading and in developmental dyslexia." *Brain and Language*, vol. 93(3), pg. 369–373, (2005). https://www.sciencedirect.com/science/article/abs/pii/S0093934X04002792?via%3Dihub.

Insert Figure 19: redrawn by the author from Chen, Zhe, and Matthew A. Wilson. "Deciphering Neural Codes of Memory during Sleep." *Trends in Neurosciences*, vol. 40(5), pg. 260–275, (2017). https://doi.org/10.1016/j.tins.2017.03.005.

Figure Page 149: from Xu, Kelvin, Jimmy Ba, Ryan Kiros, Kyunghyun Cho, Aaron Courville, Ruslan Salakhutdinov, Richard Zemel, and Yoshua Bengio. "Show, Attend and Tell: Neural Image Caption Generation with Visual Attention." ArXiv:1502.03044 [Cs], (2015). Retrieved from http://arxiv.org/abs/1502.03044.

Figure Page 157: figure composed by the author, based on graphs provided courtesy of Bruce McCandliss, from data reported in Yoncheva, Y. N., Blau, V. C., Maurer, U., & McCandliss, B. D. "Attentional Focus During Learning Impacts N170 ERP Responses to an Artificial Script." *Developmental Neuropsychology*, 35(4), 423–445 (2010). https://www.ncbi.nlm.nih.gov/pmc/articles/PMC4365954/.

Figure Page 166, top: copyright © by Stanislas Dehaene.

Figure Page 166, bottom: adapted with permission of Robert Zatorre, from data in Bermudez, Patrick, Jason P. Lerch, Alan C. Evans, and Robert J. Zatorre. "Neuroanatomical Correlates of Musicianship as Revealed by Cortical Thickness and Voxel-Based Morphometry." *Cereb Cortex*, vol. 19(7), pg. 1583–1596, (2009). https://academic.oup.com/cercor/article/19/7/1583/317010.

Figure Page 170, top: composed by the authors, based on photographs provided courtesy of György Gergely. Data from Egyed, Katalin, Ildikó Király, and György Gergely. "Communicating Shared Knowledge in Infancy." *Psychological Science*, vol. 24(7), pg. 1348–1353, (2013). https://journals.sagepub.com/doi/10.1177/0956797612471952.

Figure Page 170, bottom: composed with data from Gergely, György, Harold Bekkering, and Ildikó Király. "Rational imitation in preverbal infants." *Nature*, vol. 415(6873), pg. 755 (2002). https://www.nature.com/articles/415755a.

Figure Page 192: adapted from figure 3 in Kaplan, Frederic, and Pierre-Yres Oudeyer. "In Search of the Neural Circuits of Intrinsic Motivation." *Frontiers in Neuroscience*, 1(1), 225, (2007). https://www.frontiersin.org/articles/10.3389/neuro.01.1.1.017.2007/full. Copyright © 2007 by Kaplan and Oudeyer. This is an open-access article subject to an exclusive license agreement between the authors and the Frontiers Research Foundation, which permits unrestricted use, distribution, and reproduction in any medium, provided the original authors and source are credited. Licensed under Creative Commons Attribution License CC-BY 4.0.

Figure Page 217: copyright © by Stanislas Dehaene.

Attribution 4.0 International (CC BY 4.0) can be found at https://creativecommons.org/licenses/by/4.0/.

Attribution-ShareAlike 4.0 International (CC BY-SA 4.0) can be found at https://creativecommons.org/licenses/by-sa/4.0/.